Teaching About
SCIENTIFIC ORIGINS

Studies in the Postmodern Theory of Education

Joe L. Kincheloe and Shirley R. Steinberg
General Editors

Vol. 277

PETER LANG
New York • Washington, D.C./Baltimore • Bern
Frankfurt am Main • Berlin • Brussels • Vienna • Oxford

Teaching About
SCIENTIFIC ORIGINS

Taking Account of Creationism

Edited by Leslie S. Jones & Michael J. Reiss

PETER LANG
New York • Washington, D.C./Baltimore • Bern
Frankfurt am Main • Berlin • Brussels • Vienna • Oxford

Library of Congress Cataloging-in-Publication Data

Teaching about scientific origins: taking account of creationism /
edited by Leslie S. Jones, Michael J. Reiss.
p. cm. — (Counterpoints; v. 277)
Includes bibliographical references and index.
1. Evolution (Biology)—Study and teaching. 2. Creationism.
I. Jones, Leslie S. (Leslie Sandra) II. Reiss, Michael J. (Michael Jonathan)
QH362.T438   576.807—dc22   2006100467
ISBN 978-0-8204-7080-1
ISSN 1058-1634

Bibliographic information published by **Die Deutsche Bibliothek**.
Die Deutsche Bibliothek lists this publication in the "Deutsche
Nationalbibliografie"; detailed bibliographic data is available
on the Internet at http://dnb.ddb.de/.

Cover art by Marilyn Bechler

The paper in this book meets the guidelines for permanence and durability
of the Committee on Production Guidelines for Book Longevity
of the Council of Library Resources.

© 2007 Peter Lang Publishing, Inc., New York
29 Broadway, 18th floor, New York, NY 10006
www.peterlang.com

All rights reserved.
Reprint or reproduction, even partially, in all forms such as microfilm,
xerography, microfiche, microcard, and offset strictly prohibited.

Printed in the United States of America

To my Godparents, Freddy and John Valois, who opened my eyes to evolution by explaining marine biology and ensured that I would always be comfortable with both religion and science

—LSJ

To Mike Roberts for helping me to write better and to Sue Dale Tunnicliffe in gratitude for our work together

—MJR

Table of Contents

Acknowledgements ix

Chapter 1: Cultural Authority and the Polarized Nature of the Evolution/Creationism Controversy
- *Leslie S. Jones & Michael J. Reiss* 1

Chapter 2: The History of the Evolution/Creationism Controversy and Likely Future Developments
- *Randy Moore* 11

Chapter 3: The Warfare between Darwinism and Christianity: Who Is the Attacker and What Implications Does This Have for Education?
- *Michael Ruse* 31

Chapter 4: Capturing the Educational Potential of 'Creation Science Debates'
- *David Mercer* 43

Chapter 5: How Not to Teach the Controversy about Creationism
- *Robert Pennock* 59

Chapter 6: The Scientific Enterprise and Teaching about Creation
- *Michael Poole* 75

Chapter 7: The Theory of Evolution: Teaching the Whole Truth
- *Shaikh Abdul Mabud* 89

Chapter 8: Fundamentalist and Scientific Discourse: Beyond the War Metaphors and Rhetoric
- *Wolff-Michael Roth* 105

Chapter 9: Examining the Evolutionary Heritage of Humans
- *David L. Haury* 125

Chapter 10: Approaching the Conflict between Religion and Evolution
- *Lee Meadows* 145

Chapter 11: The Personal and the Professional in the Teaching of Evolution
- *David F. Jackson* 159

Chapter 12: Teaching for Understanding Rather than Expectation of Belief
- *Leslie S. Jones* 179

Chapter 13: Teaching about Origins in Science: Where Now?
- *Michael J. Reiss* 197

Notes on Contributors 209
Index 215

Acknowledgements

One of the more notable features of the Evolution/Creationism debate is the common illusion that there are only two diametrically opposed sides to the story. Typical discourse tends to create the impression in many people's minds that a clear delineation exists between those who support scientific theories and those who adhere to religious teachings. Epistemologically we believe that this portrayal is fallacious. Empirically most people, though the extent to which this is true differs between faith traditions, actually fall somewhere along a continuum of positions between adherence to either evolution or the creationist accounts. Rather than contributing to the polarization of religion and science, our intention has been to support science education by deciphering some of the theological and anti-scientific challenges that are at the center of this controversy. In doing so, we express our sincere appreciation to a number of scholars whose work preceded ours, establishing a growing body of literature devoted to reducing rather than exacerbating controversy over the issue of origins.

The original artwork on the cover is a drawing by Marilyn Bechler. The combination of a religious symbol with scientific figures was chosen to represent the theme of this book in which the authors address the challenges of teaching about scientific origins in the context of religious concerns. The overlapping icons portray the idea that, at least at the cultural level, there is a crucial intersection of these realms in the search for an understanding of origins. While the cross is by no means a universal representation of religion, it is certainly associated with the Christian denominations that most publicly challenge scientific explanations.

The remaining images depict a condensation of the fundamental facets of the theory of biological evolution. At the top, the molecular models stand for the biochemical constituents of living organisms that preceded the genesis of life. Various bacteria symbolize the earliest forms of life because the oldest fossils document the existence of organisms with the basic characteristics of Archaea and Eubacteria. Along the bottom of the drawing, the flower, mushroom, and primate represent the familiar Kingdoms (Plants, Fungi, and Animals). The lone paramecium just under the left arm of the cross stands for the extremely diverse Kingdom Protists into which scientists deposit every organism that does not satisfy the criteria for one of the other groups.

While neither of the series of organisms on the descending backbones of the DNA molecule should be assumed to be linear representations of evolutionary history, they both depict the documented transitions from prokaryotic ancestry through eukaryotic changes that have led to existing biodiversity. One side follows how living organisms represent the progressive advancements of key photosynthetic organisms including: cyanobacteria, algae, non-vascular and seedless vascular plants, gymnosperms, and angiosperms. The other side of the backbone traces the increasing complexity of animals: invertebrates, fish, amphibians, reptiles, and birds & mammals.

The question mark just below the top is perhaps the most important symbol in the drawing. It represents not only the universal human curiosity about origins, but also the relatively new science of Abiogenesis which is the ongoing scientific quest to decipher the mystery of how prebiotic chemicals became living organisms. The drawing is intended to emphasize the significance of the fact that science has never been able to explain the origin of life.

Cultural Authority and the Polarized Nature of the Evolution/Creationism Controversy

Leslie S. Jones & Michael J. Reiss

From the early questions of a young child asking "Where did I come from?" to the unending quest of adults who wonder "What is my purpose in life?" the primal nature of human interest in our own origins is apparent. Every culture has left evidence of the ways their people have attempted to explain origins (Maclagen, 1992). Common themes in the origin stories of different cultures demonstrate the universality of a desire to resolve the same fundamental uncertainties regarding the beginning of human existence. The elusive nature of the question of the origins of ourselves and our world lies at the heart of the clash within our culture known as the evolution/creationism controversy.

The overwhelming majority of biologists consider evolution to be the unifying concept in the biological sciences, providing a conceptual framework that links every disparate aspect of the life sciences into a single coherent discipline. Evolution is also one of the most respected and well-established tenets of any scientific field; but due frequently to organized efforts to suppress its dissemination, coverage in schools is so limited that many citizens in a range of countries do not even understand the theory, much less believe that it could be true. That such a basic aspect of scientific literacy is compromised to this extent, even in nations that take pride in their scientific and technological progress, seems particularly ironic. Aversion to evolution is much more than a historical artifact and it is no longer possible to presume that it is peculiar to conservative Protestant religions, limited to certain geographical regions, or even the United States of America.

Even though evolution is a scientific theory, it has been the center of a prolonged religious controversy. Originally, evolution was presumed to be a religious issue because of implications regarding explanations of the origin and significance of the human species. Evolution was seen as a contradiction to the accounts of origins described in the Hebrew Scriptures or what Christians refer to as the Old Testament of the Bible. The intensity of the conflict surrounding evolution in the United States was assumed to be the result of political activism by some Christians who maintain that accounts of creation in the Bible are literally true and therefore are the only correct explanations of the origin of life (Pennock, 1999). It has, however, become evident that opposition to the teaching of evolution goes beyond just religious objections and is currently being supported by a highly organized, conservative political movement. Legal challenges promoting *creation science* and more recently *intelligent design* have been used to undermine instruction about evolution and promote the inclusion of religious ideas under the guise of these ideas having scientific legitimacy.

The heart of the controversy lies in the tendency to polarize religious and scientific explanations of origins. The issue seems like an ongoing dispute that has science and religion actively battling to support the credibility of their explanations for the origin of life and human beings. The public presentation of the controversy gives the impression that biblical creationism and biological evolution are dichotomous categories referring to two mutually exclusive explanatory systems. Debate is fueled by the belief that there are only two diametrically opposed positions and a need for absolute acceptance of either evolution or the creationist accounts of origins. The absence of presentations of moderate views creates the impression in many people's minds that a clear delineation exists between those who support scientific theories and those who adhere to biblical teachings.

In reality, many, possibly most, individuals, most churches, and even most scientists are not particularly polemical and have found ways to reconcile scriptural accounts with modern scientific theories. People with moderate views avoid being caught in the conflict, moving between or combining the religious and scientific worldviews and making decisions about their own epistemologies. Those people making a narrowly scientific choice and living secular lives face little challenge and in many countries generally find their views supported in the wider culture. For people who take the religious position, a creationist stance is an immediate marker of an individual's worldview and infers an exclusive loyalty to a particular theological influence. Creationists feel compelled to defend not only the Bible, but also their own individual views which they have become convinced are incompatible with and will be actively attacked in science classrooms.

Such a polarized position obscures the fact that a large number of religious traditions find it reasonable to reconcile scientific evolution with religion. When Pope John Paul II recognized in 1996 that "evolution is more than just a hypothesis," the original Christian church of the Western World, the Roman Catholic Church, articulated an official position of acceptance. Most of the mainstream Protestant churches, including the Episcopal/Anglican, Methodist, Presbyterian, and Lutheran ones, also make no attempt to interfere with teaching about evolution and have issued explicit statements repudiating the notion of incompatibility.

The highly publicized schism between a number of religious worldviews, particularly Judeo-Christian views based on *Genesis* and mainstream Islamic readings of the *Quran*, and modern scientific explanations derived from the theory of evolution is exacerbated by the way people are asked about their views on explanations of human origins. There is a tendency to polarize religion and science in questionnaires that focus on the notion that either God created everything or God had nothing at all to do with it. The choices used in most public polls erroneously imply that scientific evolution is necessarily atheistic, coupling complete acceptance of evolution with explicit exclusion of any religious premise. Even if quantitative surveys have a middle ground, giving an option of declaring a position that combines biblical faith with evolutionary science, the language used to exemplify creationism and evolution imply a fundamental incompatibility of these explanations. As is common with quantitative social science, most surveys contain only a small number of options that makes analysis easy, 'clean' and strictly numeric. The limited number of categories forces people to codify their views to fit into, at best, three or four predetermined categories and misses essential information on the nature of what they are actually thinking.

The appearance of a polarization between scientific evolution and biblical creationism creates the impression that every individual makes a choice between religious and scientific authority. In actuality, these descriptive options are not dichotomous and individuals express personal beliefs that cover a wide range of possibilities. Eugenie Scott (1999) and others have proposed a contiguous span of views held by individuals, ranging from flat world creationists to those for whom the scientific and religions worldviews are integrated into one.

Although it is difficult to summarize the nature of religion in a way that satisfies the members of all religions, several of the characteristics of most religions (Smart, 1989; Hinnells, 1991) contribute to debate over origins. First, religions have a *practical and ritual dimension* that encompasses such elements as worship, preaching, prayer, yoga, meditation and other approaches to stilling the self. By the time students enter school, many have learned to find a comfort from this dimension that can be disrupted by scientific explanations that are so different from their existing beliefs.

Abrupt exposure to science can also disrupt the security of the *experiential and emotional dimension* of religions. At one pole are the rare visions given to some of the crucial figures in a religion's history, such as that of Arjuna in the *Bhagavadgita* and the revelation to Moses at the burning bush in *Exodus*. At the other pole are the experiences and emotions of many religious adherents, whether a once-in-a-lifetime apprehension of the transcendent or a more frequent feeling of the presence of God either in corporate worship or in the stillness of one's heart. Science, in particular evolution's connection to human origins, can seem dismissive of this dimension and may be rejected for that reason.

All religions hand down, whether orally or in writing, vital stories comprising the *narrative or mythic dimension* of their tradition. For some religious adherents such stories are believed literally, for others they are understood symbolically. In the case of the six day creation in the Judeo-Christian scriptures, scientific ideas are incongruent enough to pose cognitive challenges that teachers need to help students negotiate. Furthermore, it is crucial to recognize that creationist rhetoric receives wide circulation and some people presume that scientists will try to convince them that God was not ultimately responsible for human and cosmic origins.

The *doctrinal and philosophical dimension* of religion arises in part from the narrative/mythic dimension as theologians within a religion struggle to integrate these stories into a more general view of the world. Thus the early Christian church came to its understanding of the doctrine of the Trinity by combining the central truth of the Jewish religion – that there is but one God – with its understanding of the life and teaching of Jesus Christ and the working of the Holy Spirit. Now, modern theologians face the challenge of helping citizens integrate the significant doctrinal and philosophical teachings of religion into historically conscious worldviews that are compatible with the understandings that are the product of scientific progress.

While doctrine attempts to define the acceptable beliefs of a community of believers, the *ethical and legal dimension* regulates how believers act. So Islam has its five Pillars – Shahada (profession of faith), Salat (worship), Zakat (almsgiving), Sawm (fasting) and Hajj (pilgrimage) – while Judaism has the ten commandments and other regulations in the Torah and Buddhism its five precepts. Part of the creationist movement's objection to the theory of evolution is the perceived threat of modernism and associated immorality. Aversion to evolution also can be based on the assumption that acceptance of the theory of evolution requires atheism.

The *social and institutional dimension* of a religion relates to its corporate manifestation, for example the Sangha – the order of monks and nuns founded by the Buddha to carry on the teaching of the Dharma – in Buddhism, the umma' – the whole Muslim community – in Islam, and the Church – the communion of believers comprising the body of Christ – in Christianity. Science provides only a weaker version of this dimension through the community of peer-validated scientists. The associated loci of control involve different values than those of religion which again demand an intellectual shift that individuals with a strong religious faith may not be eager to make.

Finally, there is the *material dimension* to each religion, namely the fruits of religious belief as shown by places of worship (e.g. synagogues, temples and churches), religious artifacts (e.g. Eastern Orthodox icons and Hindu statues) and sites of special meaning (e.g. the river Ganges, Mount Fuji and Uluru [formerly known as Ayers Rock]). When evolution is regarded as a contradiction to religious tradition, there is the threat of the loss of connection to these valued sites. Thus, for some believers, embracing the explanatory value of evolutionary science means sacrificing too much in terms of the loss of the religious dimensions of their lives.

Interestingly enough, these aspects of religion also provided a number of other axes on which the relationship of science and religion can be examined. For example, the effects of the practical and ritual dimension are being investigated by scientific studies that examine such things as the efficacy of prayer and the neurological consequences of meditation; there have been a number of scientific attempts to 'explain' religious feelings; the narrative/mythic dimension of religion clearly connects, in ways that this book examines, with scientific accounts of such matters as the origins of the cosmos and the evolution of life; the doctrinal and philosophical dimension can lead to understandings that may agree or disagree with standard scientific ones (e.g. about the status of the human embryo); and the ethical and legal dimension can lead to firm views about such matters as euthanasia. Perhaps only the social, institutional and the material dimensions of religion are relatively distinct from the world of science.

There is now a very large collection of literature on the relationship between science and religion. Indeed, the journal *Zygon* specialises in this area. A frequent criticism by those who write in this area (e.g. Roszak, 1994) is of what they see as simplistic analyses of the area by those, often well known scientists, who write occasionally about it. Indeed, it is frequently argued, that the clergy both in the past and nowadays are often far more sympathetic to a standard scientific view on such matters as evolution than might be supposed (e.g. Colburn & Henriques, 2006). A thorough historical study of the relationship between science and religion is provided by John Hedley Brooke (1991). Brooke's particular aim is "to reveal something of the complexity of the relationship between science and religion as they have interacted in the past" (p.321) and it is worth quoting from his postscript at some length:

> Popular generalizations about that relationship, whether couched in terms of war or peace, simply do not stand up to serious investigation. There is no such thing as *the* relationship between science and religion. It is what different individuals and communities have made of it in a plethora of different contexts. Not only has the problematic interface between them shifted over time, but there is also a high degree of artificiality in abstracting the science and the religion of earlier centuries to see how they were related.
>
> (Brooke, 1991: 321)

Brooke's work sits alongside Barbour (1990), a classic text in the science/religion field. Barbour, in a classification that continues to prove fruitful, identified four ways in which science and religion could be seen to relate: conflict, independence, dialogue and integration.

The *conflict* model of the relationship between science and religion exists most straightforwardly when science is seen as swallowing religion. As Barbour puts it "In a fight between a boa constrictor and a wart-hog, the victor, whichever it is, swallows the vanquished" (p.4). In the UK and a number of other countries, the conflict model has recently been associated particularly with some of the writings of Richard Dawkins. A rather large body of literature is beginning to develop around Dawkins' writings on religion (see McGrath, 2005) but Dawkins' argument, and the responses to it, can fairly straightforwardly be summarised. Dawkins holds that the arguments in favour of religious faith (which he equates to a belief in God) are invalid. In particular, the argument from biological design fails because Darwinian evolution can explain even the most apparently convincing cases of design (Dawkins, 1986/1988, 1995). Dawkins also considers that religious faith is itself best seen as a sort of viral infection. The more informed theological responses to Dawkins have claimed that he either misunderstands theology or intentionally chooses not to attempt to understand it; in other words, that Dawkins is attacking a straw man.

The *independence* understanding of the relationship between science and religion sees each enterprise as having its particular worth and existing distinct from the other. This is comparable to the relationship between science and aesthetics; both might examine a building but the questions they could answer about it – 'Is it constructed safely?'; 'Is it beautiful?' – do not overlap. In Barbour's view, independence might occur because science and religion use contrasting methods or employ different languages.

When science and religion are seen in *dialogue*, there may be questions about the boundaries between them or the methods of the two fields. For example, there is literature about the extent to which certain religions facilitate or hinder the rise of science. One line of argument within the Judeo-Christian tradition has been that the orderliness of the universe is contingent rather than necessary. In other words, God could have made the universe unintelligible, thus precluding science. The fact that the universe is ordered has encouraged many scientists to feel that in studying 'the book of nature' they are attempting to understand something of the mind (or at least the workings) of God.

Finally, science and religion may be seen to be capable of *integration*. There are a number of models of integration, one of which sees science and religion contributing as partners to a comprehensive metaphysical worldview. There is, for example, a huge academic literature on process theology, an intellectual discipline that attempts to do just this. More mundanely, and somewhat closer to home for most people, many devout religious believers also accept the teachings of science and attempt, for example, to see their physical health, their feelings and the success (or otherwise) of their personal relationships as being inextricably the result both of the laws of science ('If you don't eat enough vitamins you will become unwell') and of God's laws ('For the Lord loves justice; he will not forsake his saints. The righteous shall be preserved for ever, but the children of the wicked shall be cut off', *Psalm* 37: 28).

With regard to the issue of origins, which of these four understandings of the relationship between science and religion is best depends on the precise questions being asked. If one is asking about whether dinosaurs and humans co-existed, that is manifestly a scientific question (to which the scientific answer is 'no') and any religious attempt to answer the question differently is bound to lead to conflict. If, though, one is asking about why the universe has precisely the values of the various physical constants that it does (values which, if only minutely different, would preclude the evolution of any life, let alone life sufficiently intelligent to be asking this question), then this is not really a scientific question so conflict is less likely to arise.

On the subject of origins, science and religion could appear to be engaged in the remnants of a pre-modern history of conflict, even though the modern scientific tradition 'evolved' from clerical communities in Western Europe. Many of the greatest early contributions to modern science came from thinkers like Descartes, Leibniz, and Newton who considered themselves theologians as well as scientists. However, when ideas shift public confidence at least partly from religious authorities to the scientific realm, they tend to incite a great deal of resistance and passion. As science gained credibility and popularity, new explanations of the natural world were proposed; ideas that undermined the primacy of human beings were particularly provocative. As a consequence, prominent contributors such as Galileo and Charles Darwin were accused of heresy or of leading to a corruption of morals. Such people were vilified by some sections of the religious community for their scientific ideas at the same time as other sections of the same community found their arguments acceptable, even theologically attractive.

The relationship between science and religion has changed over the years. Nevertheless, there are two key issues fuelling the evolution/creationism controversy: one to do with understandings of reality; the other to do with evidence and authority. Although it is always desperately difficult to generalize, most religions hold that reality consists of more than the observable world and many religions give weight to institutional authority in a way that science generally strives not to. For example, there is a very large religious and theological literature on the world to come. However, to labour the point, science, strictly speaking, has little or nothing to say about this question, while religious believers within a particular religion are likely to find the pronouncements on the question of even the most intelligent and spiritual of their present leaders to be of less significance than the few recorded words of their religion's founder(s).

Given the unsuccessful history of scientists' participation in educational battles over evolution, it seems hopeful that a pluralistic position, promoting cultural tolerance and individual choice, has a better chance of ensuring that students at the very least learn what evolution is. In the past, science has all too often exacerbated this evolution/creation conflict by appearing to dismiss the legitimacy of religious ideas and the validity of personal beliefs. It is impossible to deny a history that assumed the case for the *prima facie* significance of scientific knowledge and the positivist presumption that only this information was true (Cobern, 1996).

In actuality, the political side of the controversy reveals the type of fundamental epistemological and ontological questions that lie at the center of the postmodernist viewpoint. Arguments supporting the superiority of evolution on the basis of its being objective and providing rational knowledge fail to acknowledge that there are alternative valid ways of knowing the world. Thus, this dialogue regarding the intersection of scientific and religious epistemologies is well-situated in a series such as *Counterpoints: Postmodern Approaches to Education*.

It needs to be stressed that there is not a single account of how the authors in this book see the relationship between science and religion nor of how we envisage that that relationship should be taught, if it is to be taught at all. The authors of the various chapters in this book have traveled different personal and intellectual paths exploring various dimensions of science and religion. Precisely how the relationship between science and religion should be addressed in the school or college classroom is a principal focus of this book and we, as editors, make no attempt to provide a unilateral 'solution' to this 'problem'. Indeed, one of us, as editors, is a biologist trained in reproductive physiology who currently studies postsecondary science education, while the other has a doctorate in evolutionary biology and population genetics, is an academic in science education and a priest in the Church of England.

Our goal has been to assemble a collection of writings that explore the controversy from a variety of perspectives that shed light on the many facets of the evolution/creation controversy. Our hope is that the varied contributions will help those who are responsible for teaching science as well as those who, more generally, are concerned at how science can be presented in a way that is true to itself and also respects those with a religious faith.

In Chapter 2, biologist Randy Moore examines the legal history of the creationism/evolution controversy and its likely future developments. He traces the way in which creationists in the USA have lost just about every legal battle over the teaching of evolution in schools and shows how the 'battle' (the legal system almost demands a conflict approach) has resulted in attempts by creationists to get anti-evolution policies in state educational guidelines and to introduce intelligent design into the classroom.

In Chapter 3, philosopher Michael Ruse examines the 'warfare' between Darwinism and Christianity. Maintaining that it is possible now to have a happy synthesis between science and religion, he criticizes Richard Dawkins and contrasts Dawkins' approach to the science/religion issue with those of other leading evolutionary scientists such as Stephen Jay Gould and Simon Conway Morris.

In Chapter 4, David Mercer, an academic in science and technology studies, considers the way the creation science debate often encourages a discourse that is pre-occupied with attempting to resolve larger debates in epistemology involved with identifying and describing the philosophical essences of science and religion. He concludes by sketching some of the ways science education and the popularization of science could be improved by paying more attention to science as local knowledge and practice.

In Chapter 5, historian Robert Pennock presents a fine-grained analysis of the controversy as enacted in Kansas. He examines and rejects the claims of intelligent design creationists that they have a right of academic freedom to see their case presented in schools that, in their view, passes the constitutionality test, that it is supported by a federal mandate, that opinion polls support the notion, and that doing so is good pedagogy. Instead he concludes that only 'real science' should be taught in science classes.

In Chapter 6, science educator Michael Poole argues that teaching about Creation, in its theological sense, and a professional dedication to the scientific enterprise can with integrity go hand in hand. Pointing out that while science studies are limited to the natural world it does not follow that therefore there is nothing other than the natural world, he holds that scientific explanations of the mechanisms of the world are logically compatible with religious explanations of divine agency and purpose.

In Chapter 7, Abdul Mabud, a scientist by training and now Director General of the Islamic Academy in the UK, while not presenting an argument in support either of the theory of evolution or of its critics, argues that the presentation of the origins of the various forms of life in school textbooks and lessons is highly biased in favour of the theory of evolution and points out that the scientific findings contrary to the evolutionary perspective are hardly mentioned. He concludes that educators should provide a balanced view of evolution, by presenting to students a careful and fair evaluation of the points both for and against the theory.

In Chapter 8, science educator Wolff-Michael Roth begins by noting that whereas students' everyday (non-scientific) descriptions of physical events have been well studied over the past quarter of a century, other experiential sources and discourses including ethical, religious, and emotional issues have been less studied. Drawing on extensive data from one student, he examines how language is used to manage the relationship between science and religion and discusses how teachers should respond to students with strong religious views.

In Chapter 9, science educator David Haury begins by presenting a case for the study of human evolution in the school curriculum. He maintains that teaching about human evolution greatly enriches science content, responds to fundamental curiosities and concerns that people have, provides a personally meaningful context for examining scientific ways of knowing, and informs decision making in an era of powerful biotechnologies. He concludes by presenting a response to those who oppose the teaching of human evolution.

In Chapter 10, Lee Meadows, who came to science education from a very conservative religious tradition, sees evolution and religion as two different subjects that cannot be resolved. Indeed, he feels that when biology teachers approach the conflict with a resolution mentality, taking as theirs the responsibility to cause students to resolve the conflict between the two clashing worldviews, they can do more harm than good. Instead he presents conflict management as a way to help students study evolution when their religious beliefs create conflict with evolution.

In Chapter 11, former earth science teacher David Jackson writes autobiographically and describes how he teaches about cosmological evolution and 'The Creation Controversy' in his science methods course for prospective middle school teachers, trying to nudge them towards engaging in experiences that give them the opportunity better to understand evolution as an important exemplar of science in general.

In Chapter 12, Leslie S. Jones describes how she learned to recognize the intensity of creationist aversion to evolution and now considers this issue too significant to ignore in college biology courses. She cautions that for individuals whose initial explanation of human origins came out of a strong religious tradition, scientific explanations can seem radically incongruent. However, science teachers can do a great deal to minimize resistance and encourage creationists to at least learn what evolution actually is.

Finally, in Chapter 13 Michael Reiss attempts to draw together the threads of the book, clarify areas of agreement and areas of disagreement and suggest possible ways forward. He organizes his chapter around three main themes: teaching about the nature of knowledge, teaching about controversial issues, and teaching about matters of personal significance. He concludes that the strongest argument as to why science teachers should deal with the relationship between science and religion when teaching about origins, whether in biology, earth sciences or astronomy, is that it is good science teaching so to do, and discusses ways in which this can be done.

References

Barbour, I. G. (1990). *Religion in an age of science: The Gifford Lectures 1989-1991, volume 1*. London: SCM.
Brooke, J. H. (1991). *Science and religion: Some historical perspectives*. Cambridge: Cambridge University Press.
Cobern, W. W. (1996). Worldview theory and conceptual change in science education. *Science Education, 80*, 579-610.
Colburn, A. & Henriques, L. (2006). Clergy views on evolution, creationism, science, and religion. *Journal of Research in Science Teaching, 43*, 419-442.
Dawkins, R. (1986/1988). *The blind watchmaker*. London: Penguin.
Dawkins, R. (1995). *River out of Eden: A Darwinian view of life*. London: Weidenfeld & Nicolson.
Hinnells, J. R. (1991). *A handbook of living religions*. London: Penguin Books.
McGrath, A. (2005). *Dawkins' God: Genes, memes, and the meaning of life*. Oxford: Blackwell.
Maclagan, D. (1992). *Creation myths: Man's introduction to the living world*. New York: Thames and Hudson.
Pennock, R. T. (1999). *Tower of Babel: The evidence against the new creationism*. Cambridge, MA: MIT Press.
Roszak, T. (1994, March). God and the final frontier, *New Scientist*, 28, 40-41.
Scott, E. (1999). The creation/evolution continuum. *Reports of the National Center for Science Education, 19*, 16-23.
Smart, N. (1989). *The world's religions: Old traditions and modern transformations*. Cambridge: Cambridge University Press.

The History of the Evolution/Creationism Controversy and Likely Future Developments

Randy Moore

The publication of Charles Darwin's monumental *On the Origin of Species* in 1859 was greeted with a variety of responses. Most scientists accepted or embraced—albeit sometimes reluctantly—Darwin's brilliant idea. For example, Darwin's advocate Thomas Huxley summarized the responses of many scientists when he proclaimed, "How extremely stupid [of me] not to have thought of that!" However, Darwin's *On the Origin of Species* elicited a much different response from many theologians and religious leaders. Darwin's idea threatened these and other people, for they believed that Darwin's idea:

- removed the necessity of a deity to account for life's diversity.
- provided a different and testable place for humans in nature. Whereas many people believed that humans were created specially by a deity, Darwin claimed that humans evolved from animals.
- challenged the common Victorian view that the natural world was harmonious, purposeful, and altruistic. In Darwin's world, organisms were in a brutal, amoral, and violent struggle to survive and reproduce.

Although Darwin never actively promoted his book, others did, both in the UK and the US. In England, Huxley became "Darwin's bulldog," and, in the US, Harvard's Asa Gray—an evangelical Christian and arguably one of the country's leading scientists—eased Protestants' fears by announcing that Darwin's idea did not necessarily conflict with Christianity. By the 1890s, there was relatively little controversy associated with Darwin's views among scientists; by then, Darwin's model for evolution by natural selection was featured prominently in virtually all best-selling high school biology textbooks (Moore, 2001b).

However, in the early 1900s, society and societal views of evolution began to change. In 1906, George McReady Price, whose ideas would be transformed in the 1960s into "creation science," denounced evolution as "a most gigantic hoax." Soon thereafter, fundamentalist leader William Bell Riley proclaimed that evolution gives humans "a slime sink for origin and an animal ancestry," whereas Christianity makes humans "the creature and child of the most high." Many people throughout the US, especially in the South, began to wonder whether America's encroaching "modernism" had lost its moral compass as crime rates increased, sexual morays changed, people began abandoning rural life for large cities, and the country entered World War I. Many people began to long for the "old time religion" that they remembered from earlier times when life was seemingly simpler and more secure. These concerns about America's moral well-being became explicit with the publication of *The Fundamentals*, a series of small books that outlined how the US should use biblical literalism to regain its prewar innocence and prosperity. As fundamentalism became increasingly popular, fundamentalists such as William "Billy" Sunday, John Roach Straton, J. Frank Norris, William Bell Riley, Aimee Semple McPherson, and William Jennings Bryan[1] preached throughout the South and elsewhere about the evils of evolution. These leaders of fundamentalism believed that the teaching of evolution was equivalent to the teaching of atheism, and was directly responsible for a variety of social problems, including war, crime, and declining morals. To these leaders, saving the nation's soul required that the country return to a Bible-based faith that left little or no room for Darwin's ideas. In simplest terms, the teaching of evolution had to be stopped.

Fundamentalists' attacks on evolution were often harsh, and usually focused on the teaching of evolution in public schools:[2] For example:

- Sunday and McPherson used theatrical services to link evolution with eugenics, prostitution, and crime.
- Straton, known as "The Fundamentalist Pope," told his followers that it was "better to wipe out all the schools than … permit the teaching of evolution."
- Norris, a fiery Texan who led the nation's largest Baptist church, described evolution as "that hell-born, Bible-destroying, German rationalism" as he campaigned to have evolutionists fired from their teaching positions at colleges and universities.

Leaders of the fundamentalist movement soon realized that their antievolution attacks had a strong appeal to the public; by the 1920s, all the leaders had used their antievolution message to gather large followings. For example, William Jennings Bryan's speeches often attracted crowds of thousands, and Billy Sunday's 18-day revival in Memphis in early 1925—at which he declared Darwin an "infidel" and claimed that public education was "chained to the devil's throne"—attracted more than 10% of Tennessee's residents. William Bell Riley's World's Christian Fundamentals Association (WCFA), which was founded in 1919,[3] sponsored meetings that focused on banning the teaching of evolution and were attended by thousands of believers. Throughout the early

---

[1] Bryan was a three-time Democratic nominee for president of the United States, and a former Secretary of State. Bryan, who had electrified the 1896 Democratic National Convention with his famous "Cross of Gold" speech, argued that if evolution were true, humans could not overcome their animal nature, thereby destroying all hopes for societal reform.

[2] By 1920, enrollment in U.S. high schools reached 2.5 million students (32% of 14-to-17-year-olds in the country).

[3] The inaugural meeting of the WCFA in Philadelphia was attended by 6000 people, and was described by Riley as "an event of more historical moment than the nailing, at Wittenberg, of Martin Luther's ninety-five theses."

1920s, Riley used the WCFA to organize efforts in many state legislatures to ban the teaching of evolution.

The first legislative vote to ban the teaching of evolution occurred on March 9, 1922, in Kentucky, just weeks after a Kentucky teacher was fired for teaching that the Earth is round. After being told by William Jennings Bryan that "the movement [to ban the teaching of evolution] will sweep the country and will drive Darwinism from our schools," the state legislature defeated the proposed law by a vote of 42–41.[4] Later in 1922, the Southern Baptists (who were usually the leaders of fundamentalism and the antievolution movement in the South) reiterated their belief that the Bible and evolution were irreconcilable by proclaiming that "no man can rightly understand evolution's claim as set forth in the textbooks of today, and at the same time understand the Bible." At their convention the following year, the Southern Baptists instructed scientists to concede the authority of the Bible, including the Virgin Birth and the physical Resurrection. At about the same time, the Arkansas Baptist Convention instructed the state's Baptist institutions not to employ anyone who endorsed evolution, and the Oklahoma Baptist Convention demanded that the state ban the teaching of evolution. On March 24, 1923, the Oklahoma Baptists got what they wanted when Oklahoma governor John Walton signed the nation's first antievolution law. That law, which was suggested by William Jennings Bryan, offered free textbooks to public schools whose teachers would not mention evolution, and banned the use of textbooks that promoted "a materialistic conception of history, that is, the Darwin theory of evolution versus the Bible theory of creation." Although this was the nation's first antievolution law, it was passed with little fanfare and went largely unnoticed.[5]

Throughout the South, fundamentalists continued to demand that state legislatures ban the teaching of evolution. In Tennessee, state legislator John Butler drafted legislation to ban the teaching of evolution and "protect our children from infidelity." Contrary to its official title, Butler's proposal (House Bill 185: "An Act Prohibiting the Teaching of the Evolutionary Theory …") did not ban the teaching of all of evolution; instead, it banned only the teaching of human evolution. His legislation encountered virtually no opposition, and Tennessee governor Austin Peay ("The Maker of Modern Tennessee") signed the bill into law in early 1925 because of his "deep and widespread belief that something is shaking the fundamentals of the country … [and that] an abandonment of the old-fashioned faith and belief in the Bible is our trouble." Butler's legislation, thereafter known as the Butler Law, became the most

---

[4] Seventy-seven years later, Kentucky deleted the word *evolution* from its state educational guidelines.
[5] Oklahoma's law was repealed in 1925 after Walton was impeached.

famous of all antievolution laws, and served as a model for other states' attempts to pass bans on the teaching of evolution.[6]

When the fledgling American Civil Liberties Union (ACLU) in New York learned of Tennessee's new law, it placed advertisements in several Tennessee newspapers "looking for a Tennessee teacher who is willing to accept our services in testing [the Butler] law in the courts." George Rappleyea, a businessman in Dayton, Tennessee, saw the ACLU's advertisement and, believing that a trial to test the law could stimulate Dayton's struggling economy, discussed the advertisement with other local businessmen. After agreeing that their tiny town could profit by sponsoring a high-profile test of the law, they asked the local high school's biology teacher to participate in the proposed trial. When he refused, they asked coach and substitute science teacher John Scopes to be their "guinea pig" and test the law. Scopes agreed, believing that little would come from the trial. He was wrong; the so-called "Scopes Monkey Trial" that resulted from his arrest and trial became the first "trial of the century" and one of the most famous and enduring events in American history.

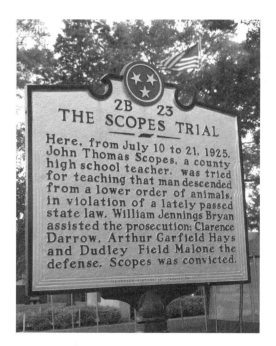

Figure 2.1 Historic Marker at the Site of the Trial

---

[6] The Butler Law would not be repealed until 1967, when Tennessee faced an expensive and embarrassing lawsuit filed by Gary Scott, a Campbell County (Tennessee) biology teacher who had been fired for teaching evolution.

The Trial of the Century: Can a State Ban the Teaching of Evolution?

William Bell Riley had campaigned for the passage of Tennessee's ban on the teaching of evolution. When he heard that Scopes would challenge the law, he quickly mobilized the WCFA to aid the state's cause. As Riley and the WCFA denounced "the teaching of the unscientific, anti-Christian, atheistic, anarchistic, pagan rationalistic evolutionary theory," Riley asked his close friend and the WCFA's most famous ally, William Jennings Bryan, to represent the WCFA at the trial and help prosecute Scopes in Dayton. Realizing the importance of the trial, Bryan quickly agreed, and began a new assault on the teaching of evolution, denouncing it as "poison" and blaming it for World War I and "all the ills from which America suffers." Bryan's involvement made the trial a world-class event, and quickly prompted famed defense attorney Clarence Darrow to volunteer his services for Scopes' defense.[7] As organizations such as the Tennessee Academy of Science and the National Education Association remained silent, Bryan and Darrow began using the media to promote the trial, with Bryan finally claiming that the trial would be "a contest between evolution and Christianity ... a duel to the death ... the two cannot stand together."

Scopes' trial took place in July 1925 in Dayton, Tennessee. It was, by all accounts, a sensational event. Newspaper reporters came from all over the world to cover the trial, and thousands of spectators flooded tiny Dayton. The once-sleepy rural town soon had a carnival-like atmosphere, with preachers and salespeople on every corner shouting their messages to the crowds. It was, as famed journalist H. L. Mencken wrote, "a religious orgy."

The Scopes Trial reached a climax on July 20 when Darrow shocked everyone by calling William Jennings Bryan (one of Scopes' prosecutors) to the witness stand as an expert on the Bible. Darrow's two hours of rapid-fire questions finally resulted in Bryan admitting that he did not believe that everything in the Bible is literally true. The next day, after deliberating only nine minutes, the jury found Scopes guilty of the misdemeanor of teaching human evolution. The judge fined Scopes $100.

Bryan was proud of his performance at the trial; on the Saturday after the trial, he told reporters that "If I should die tomorrow, I should feel that much has been accomplished in the greatest cause for enlightening humanity ever known. I believe that on the basis of the accomplishments of the past few weeks, I could truthfully say, 'Well done.'" Ironically, Bryan did die the next day—becoming a martyr for the cause—and was buried five days later in Arlington National Cemetery under the tiny epitaph "He Kept the Faith."

Darrow appealed Scopes' conviction; he wanted the case to go to the US Supreme Court. However, in 1927 the Tennessee Supreme Court upheld the constitutionality of the Butler law, but reversed Scopes' conviction on a

---

[7] Earlier in July of 1925, Scopes' sister Lela (a mathematics teacher in Paducah, Kentucky) was told that her "services will not be desired unless she renounces the theory of evolution." When Lela told the school board that she endorsed her brother's views "in their entirety," she was fired.

technicality. Darrow had saved his now-famous client from having to pay the $100 fine, but there was now no way to continue his appeal of Scopes' famous case. *John Thomas Scopes v. State of Tennessee* was over. Following Tennessee's lead, Mississippi and Arkansas also passed laws banning the teaching of evolution, but similar efforts elsewhere failed. The ACLU looked for teachers to challenge these laws, but no one volunteered.[8]

Although the Scopes Trial accomplished nothing legally, it had a major impact on society. After the Scopes Trial, the word *evolution* disappeared from US biology textbooks, and most biology teachers no longer taught the subject in their classrooms. As fundamentalist preacher George McReady Price (1929) observed, "Virtually all textbooks on the market have been revised to meet the needs of fundamentalists." Bans on teaching evolution in the public schools of Tennessee, Arkansas, and Mississippi remained in effect and unchallenged for more than 40 years.

Sputnik and the Reemergence of the Teaching of Evolution

The event that eventually put the evolution/creationism controversy back in the US court system came from an unlikely source—the former Soviet Union. On October 4, 1957, the Soviet Union successfully launched *Sputnik I*, the first orbiting artificial satellite, and triggered widespread concern that the US was lagging behind the Soviet Union in science and technology. This concern soon led to political action. In 1958, Congress passed the National Defense Education Act, which encouraged the National Science Foundation to develop state-of-the-art science textbooks. Early the following year, the Biological Sciences Curriculum Study (BSCS) was created and, with the help of the country's top biologists, began working on a new kind of biology textbook for public schools. The resulting books emphasized evolution as the foundation of biology, and quickly "put evolution back in the biology classroom."

But before the BSCS could complete its books, Henry Morris and John Whitcomb published an influential and popular book entitled *The Genesis Flood: The Biblical Record and its Scientific Implications*. *The Genesis Flood* used flood geology as the central paradigm of creationism while claiming that humans lived with dinosaurs and that natural phenomena can be explained by principles of biblical inerrancy (e.g., "the fossil record, no less than the present taxonomic classification system and the nature of genetic mutation mechanism, shows exactly what the Bible teaches"). Morris and Whitcomb reprimanded scientists for studying

---

[8] The local high school offered Scopes a new contract if he would adhere "to the spirit of the evolution law," but Scopes instead decided to enroll in graduate school at the University of Chicago on a scholarship funded by scientists and reporters who attended his trial. Scopes spent most of the rest of his life working as a geologist in South America, refusing lucrative offers to cash in on his accidental fame. He never again was involved significantly in the evolution/creationism controversy. He died in 1970 and was buried in a family plot in Paducah, Kentucky under the epitaph "A Man of Courage."

the details of creation,⁹ and repackaged George McReady Price's earlier ideas about the Bible and science to create what became known as "creation science." Soon thereafter, the US Supreme Court banned state-sponsored prayer (*Engel v. Vitale*) and Bible reading (*School District of Abington Township v. Schempp*), and Morris formed the Christians-only Creation Research Society to organize believers of "creation science" into a national organization. These events, combined with the release in 1960 of the popular and award-winning movie *Inherit the Wind* (a fictitious account of the Scopes Trial meant to be a commentary on the anticommunist hysteria of the McCarthy era), returned the evolution/creationism controversy to the public's attention.¹⁰

The BSCS books were published in 1963 and were an immediate hit. Unlike most other biology textbooks, which either ignored evolution or presented it apologetically, the BSCS books were based on evolution. Biology teachers throughout the country, anxious for such an approach, liked the books; within a few years, the BSCS books were used in almost half of the nation's biology classrooms. Not everyone, however, was pleased with the books. When students and parents realized what the books contained, there were a variety of protests, and several religious organizations and individuals began campaigns to exclude the BSCS books from their schools' biology classrooms.

Meanwhile, in Arkansas, a young teacher named Susan Epperson faced a dilemma. As a public school teacher, she believed that she was a role model and should obey the law. However, Susan was also a biology teacher who wanted to teach an up-to-date course. To Susan, this meant that she had to base her course on evolution. To resolve this dilemma, she decided in December 1965 to ask the courts to determine if the Arkansas law banning the teaching of evolution was constitutional. Epperson, an evangelical Christian and Sunday school teacher, understood that her challenge of the Arkansas ban on teaching evolution could be controversial. To avoid this, Epperson released the following statement on the day that her court challenge was announced:

> I am a teacher by profession. I chose to become a teacher because I believe that teaching is the most important profession to which a person can dedicate his talents and energies. I pursued an education course in college to become a science teacher as competent as my capabilities will permit. I received a bachelor's degree from the College of the Ozarks and a Master of Science degree from the University of Illinois.
>
> My mother is a public school librarian and my father has been a professor of biology for almost half a century. They are both dedicated Christians who see no conflict between their belief in God and the scientific search for truth. I share this belief.
>
> As a new teacher (this is my second year), and, as a native Arkansan, Arkansas' anti-evolution law has disturbed me more than a little.
>
> I do not try to teach my students what to think. I try to teach them how to think, how to make sound judgments about the various relevant alternatives. In doing this, it

---

⁹ According to Morris and Whitcomb, "the fully instructed Christian knows that the evidence for full divine inspiration of Scripture is far weightier than the evidences for any fact of science."

¹⁰ The world premier for *Inherit the Wind* was at Dayton, Tennessee, and was attended by John Scopes, who was honored with a parade and given a key to the city.

is my responsibility to expose my students to and encourage them to seek after as much of the accumulated scientific knowledge and theories as possible. Rational knowledge, as accurate and balanced as humanly possible at any given time, is essential to the making of sound value judgments.

When [I was asked] to become the plaintiff in this test suit I agreed to do so because of my concept of my responsibilities both as a teacher of biology and as an American citizen. This law, prohibiting any teacher from discussing in any way the Darwinian theory of evolution, compels me either to neglect my responsibility as a teacher or to violate my responsibility as a citizen. As a responsible biology teacher, it is my duty to discuss with my students and to explain to them various scientific theories and hypotheses in order that they may be as educated and enlightened as possible about matters pertaining to science, including the theories of Darwin as set forth in "On the Origin of Species" and in "The Descent of Man." However, when I do this, I become an irresponsible citizen—a law violator—a criminal subject to fine and dismissal from my job. On the other hand, if I obey the law, I neglect the obligations of a responsible teacher of biology. This is the sure path to the perpetuation of ignorance, prejudice, and bigotry.

The only recourse available to me which is consistent with my concept of my responsibilities as a teacher and citizen is this test suit. It is my fervent hope that this suit will resolve this dilemma not only for me but also for all other Arkansas teachers.

Susan's case created an immediate sensation; throughout the country, it was referred to as "Scopes II." Despite her statement to the contrary, Susan was labeled a Communist, infidel, and atheist by people opposed to the teaching of evolution. Susan won her initial challenge when an Arkansas chancery court ruled on May 27, 1967, that the Arkansas law banning the teaching of evolution was unconstitutional. However, when this decision was overturned by a one-paragraph opinion issued by the Arkansas Supreme Court less than two weeks later, Epperson appealed her case to the US Supreme Court. On November 12, 1968, the US Supreme Court ruled unanimously in *Epperson v. Arkansas* that it is unconstitutional to ban the teaching of evolution in public schools.

As a result of *Epperson v. Arkansas*, all of the laws banning the teaching of evolution had been overturned by 1970. However, as we'll see in the following section of this chapter, there would be many other court challenges to thwart the teaching of evolution.

Following Susan to the Courthouse: *Epperson v. Arkansas*

During the 1960s, the evolution/creationism debate surfaced periodically throughout the US. A variety of states considered bills requiring that creationism and evolution be given "equal time," and public sentiment remained strongly in favor of teaching creationism in public schools. For example, a poll in 1972 of students at Rhea County High School (the school at which John Scopes had taught) found that three-fourths of the students were creationists who believed that evolution produces corruption, lust, greed, drug addiction, war, and genocide.

Throughout the US, individuals and groups who felt that their rights were being abridged began to seek relief in the US court system. Protests and lawsuits for the rights of women, minorities, and other groups became increasingly common. Not surprisingly, opponents of the teaching of evolution tried to use the same strategy to promote their goals. These lawsuits answered a variety of questions about the teaching of evolution and creationism in public schools.

In the following pages, I answer the major questions that US courts have addressed regarding the evolution/creationism controversy. Details regarding each case are presented elsewhere (Moore, 2002a).

If a student claims that evolution offends and is incompatible with their religious beliefs, must teachers modify their teaching to accommodate the student's right to religious freedom?

No. *Wright v. Houston Independent School District* (1973)—the first lawsuit to be initiated by creationists—was dismissed when the Fifth Circuit Court of Appeals ruled that to be insulated from scientific findings incompatible with one's religious beliefs does not accompany the free exercise of religion. As had been noted in *Epperson v. Arkansas* (1968), "there is and can be no doubt that the First Amendment does not permit the State to require that teaching and learning must be tailored to the principles or prohibitions of any religious sect or dogma" (p. 106), and that there can be no legitimate state interest in protecting particular religions from scientific views "distasteful to them." Citing *Epperson*, the Fifth Circuit Court of Appeals ruled in *Wright v. Houston Independent School District* (1973) that "Federal courts cannot by judicial decree do that which the Supreme Court has declared the state legislatures powerless to do, i.e., prevent teaching the theory of evolution in public schools for religious reasons" (p. 137).

Science and religion necessarily deal with many of the same questions, and they may often provide conflicting answers. But as the US Supreme Court noted in 1952 in *Burstyn v. Wilson*, it is not the business of government to suppress real or imagined attacks on a particular religious doctrine. Science teachers in public schools should not be expected to avoid discussions of every scientific issue on which a religion claims expertise.

Textbooks produced in the 1960s by the Biological Sciences Curriculum Study promoted the teaching of evolution and were subsidized by the federal government. If the government uses tax money to produce science textbooks that promote evolution, must it also provide funds to produce textbooks that promote creationism? Can the government use tax money to promote the teaching of evolution?

In 1972, William Willoughby—acting "in the interests of forty million evangelistic Christians" in the US—sued the director of the National Science Foundation and others for funding the popular and decidedly pro-evolution textbooks produced in the 1960s by the Biological Sciences Curriculum Study (BSCS). Willoughby's lawsuit (*Willoughby v. Stever*, 1972) was dismissed by the D.C. Circuit Court of Appeals, on the grounds that (1) the BSCS books disseminated scientific findings, not religion, (2) governmental agencies such as the National Science Foundation can use tax money to disseminate scientific findings, including those related to evolution, and (3) publicly funded science textbooks cannot be tailored to particular religious beliefs. Similarly, a Georgia court ruled recently that a textbook did not violate the Establishment or Free Exercise Clauses when the text stated "creationism is not a scientific theorem capable of being proven or disproven through scientific methods" (*Moeller v. Schrenko*, 2001, p. 201).

Can science teachers teach creationism if their school district adopts a course textbook that promotes creationism?

No. Indiana Superior Court Judge Michael T. Dugan ruled in *Hendren v. Campbell* (1977) that creationism-based biology textbooks such as *Biology: A Search for Order in Complexity* advance a specific religious perspective, and that the use of such books would ensure "the prospect of biology teachers and students alike being forced to answer and respond to continued demand for correct fundamentalist Christian doctrine in public schools." It is unconstitutional to adopt a science textbook that promotes creationism because such textbooks are sectarian in content and entangle the state with religion.

If the government uses public funds to produce public exhibits that promote evolution, must it also provide funds to produce public exhibits that promote creationism?

No. In *Crowley v. Smithsonian Institution* (1980), attempts to ban federally funded evolution-based exhibits were dismissed on grounds similar to those used to dismiss *Willoughby v. Stever* (1972). The D.C. Circuit Court of Appeals determined that using public funds to promote evolution was constitutional and that the government is not required to provide public funds to promote, or give equal time to, creationism.

Must science teachers who teach evolution give "equal time" to creationism?

In *McLean v. Arkansas Board of Education* (1982), Federal Judge William Overton ruled that Arkansas' law requiring "equal time" for creationism was unconstitutional (as per the Lemon Test in *Lemon v. Kurtzman*, 1971), was based on "an inescapable religiosity," and "advanced particular religious beliefs" (pp. 12, 13). In 1987, the US Supreme Court ruled (by a majority of 7–2) in *Edwards v. Aguillard* that Louisiana's law requiring "balanced treatment" for creationism was also unconstitutional and "facially invalid." The Court was particularly concerned with the impact of the "Creationism Act" on impressionable children in public schools having mandatory attendance. These cases, combined with *Daniel v. Waters* (1975) (which overturned Tennessee's "Genesis Act" requiring public schools to give equal emphasis to evolution and the Genesis version of creation), doomed future attempts by legislatures to require "equal time" and "balanced treatment" for creationism in public schools.

Does "creation science" have any educational merit as science?

No. In *McLean v. Arkansas Board of Education* (1982), Federal Judge William Overton ruled that "creation science is simply not science," that "creation science is not guided by natural law," and that "creation science has no scientific merit or educational value as science" (p. 22). Instead of being scientists, Overton ruled, creationists "take the literal wording of the Book of Genesis and attempt to find scientific support for it ... A theory that is by its own terms dogmatic, absolutist, and never subject to revision is not a scientific theory" (p. 17).[11] As Overton noted, "if the unifying idea of supernatural creation by God is removed [from the law], the remaining parts [of the law] explain nothing and are meaningless assertions" (p. 15).

If students, their parents, and the local taxpayers want teachers to teach creationism as well as evolution, can the teachers teach them both? Isn't teaching both creationism and evolution in science classes only fair?

There are numerous problems with this logic. For example, there are thousands of different creation stories. Since the Constitution requires that schools must be religiously neutral, a teacher cannot present any particular creation story as being more "true" than others. As noted by the Fifth Circuit Court of Appeals in *Wright* (1973), "to require the teaching of every theory of human origin ... would be an unwarranted intrusion into the authority of public school systems to control the academic curriculum" (p. 137).

The popularity of creationism is irrelevant to it being taught in public schools. As noted by Judge Overton in *McLean v. Arkansas Board of Education* (1982), "the application and content of the First Amendment principles are not determined by public opinion polls or by a majority vote. Whether the proponents of [teaching creationism] constitute the majority or the minority is quite irrelevant under a constitutional system of government. No group, no matter how large or small, may use the organs of government, of which the public schools are the most conspicuous and influential, to foist its religious beliefs on others" (p. 26).

Courts have often been required to invalidate statutes that advance religion in public elementary and secondary schools, even though a majority supports the laws, as exemplified by *Wallace v. Jaffree* (1985) (Alabama law authorizing a moment of silence for school prayer), *Stone v. Graham* (1980) (posting a copy of the Ten Commandments on walls of public schools), *Abington School District v.*

---

[11] Creationists' dogmatism is explicit in statements such as "by definition, no apparent, perceived, or claimed evidence in any field ... can be valid if it contradicts the scriptural record" (Answers in Genesis), all of the Bible's "assertions are historically and scientifically true" (*Creation Research Society Quarterly*), "The final and conclusive evidence against evolution is the fact that the Bible denies it ... If the Bible teaches it, that settles it, whatever scientists might say, because it's the word of God" (Henry Morris), and "I want you to have all the academic freedom you want, as long as you wind up saying the Bible account [of creation] is true and all others are not" (Jerry Falwell; see discussion in Moore, 2000).

Schempp (1963) (daily reading of Bible), and *Engel v. Vitale* (1962) (recitation of denominationally neutral prayer).

It is unconstitutional to teach creationism in science classes of public schools (*Edwards v. Aguillard*, 1987). One reason is because children are highly impressionable. The Supreme Court has determined that teachers must not inculcate religion; school administrators have a duty to make certain that public school teachers do not inculcate religion (see also, *Lemon v. Kurtzman*, 1971 and *Hellend v. South Bend Community School Corporation*, 1996).

Creation science is not science (*McLean v. Arkansas Board of Education*, 1982), and evolution is discredited if it must be counterbalanced by the teaching of creationism (*Edwards v. Aguillard*, 1987).

All citizens of the US have a First Amendment right to free speech. Doesn't this right to free speech entitle teachers to teach creationism in science classes of public schools?

No. In *Webster v. New Lenox School District #122* (1990), the Seventh Circuit Court of Appeals ruled that a teacher does not have a First Amendment right to teach creationism in a public school. A school can direct a teacher to "refrain from expressions of religious viewpoints in the classroom and like settings" (*Bishop v. Aronov*, 1991, p. 1077). In 1996, the Seventh Circuit upheld the firing of a teacher, noting that the teacher did not have a First Amendment right to teach creationism, and, in fact, the school had a constitutional duty to make certain, given the Religion Clauses, that subsidized teachers did not inculcate religion (*Hellend v. South Bend Community School Corporation*).

Can a school district force a science teacher to stop teaching creationism? If the teacher refuses to teach evolution, can the teacher be reassigned?

Yes. In *Webster v. New Lenox School District #122* (1990), the Seventh Circuit Court of Appeals ruled that a school district can ban a science teacher from teaching creationism. As noted in *Edwards v. Aguillard* (1987) and *Peloza v. Capistrano Unified School District* (1994), "the Supreme Court has held unequivocally that while the belief in a divine creator of the universe is a religious belief, the scientific theory that higher forms of life evolved from lower forms is not" (p. 521).

Can a school require that a teacher teach evolution? If so, doesn't this violate a teacher's right to free speech?

In *Peloza v. Capistrano Unified School District*, the Ninth Circuit Court of Appeals ruled in 1994 that requiring a science teacher to teach evolution is not a violation of the Establishment Clause of the US Constitution. In dismissing Peloza's lawsuit, the court noted that "since the evolutionist theory is not a religion, to require an instructor to teach this theory is not a violation of the Establishment Clause … Evolution is a scientific theory based on the gathering and studying of data, and modification of new data. It is an established scientific theory which is used as the basis for many areas of science. As scientific methods advance and become more accurate, the scientific community will revise the accepted theory to a more accurate explanation of life's origins. Plaintiff's assertions that the teaching of evolution would be a violation of the Establishment Clause is unfounded" (pp. 521–522).

Hasn't the US Supreme Court endorsed the teaching of "evidence against evolution"?

No. However, in the minority opinion of *Edwards v. Aguillard* (1987), US Supreme Court Justice Scalia argued that "the people of Louisiana, including those who are Christian fundamentalists, are quite entitled, as a secular matter, to have whatever scientific evidence there may be against evolution presented in their schools, just as Mr. Scopes was entitled to present whatever scientific evidence there was for it." Although the majority of the Court concluded that Louisiana's antievolution law was unconstitutional, it also noted that "we do not imply that a legislature could never require that scientific critiques of prevailing scientific theories be taught" and that "teaching a variety of scientific theories about the origins of humankind to school children might be validly done with the clear secular intent of enhancing the effectiveness of science instruction."

Does a science teacher's right to free speech entitle him or her to teach the alleged "evidence against evolution" that Justice Scalia cited?

No. In the late 1990s, biology teacher (and creationist) Rodney LeVake taught the alleged evidence against evolution to his students because he believed that evolution is impossible and that there is no evidence to show that it actually occurred. When LeVake was reassigned, he sued. In *LeVake v. Independent School District #656* (2000), the Minnesota Court of Appeals ruled that "a school board's decision to assign a public school teacher to teach a different class because the teacher refused to teach his former assigned class according to the curriculum established by the school board did not violate the teacher's right to free exercise of religion. A public school teacher's right to free speech as a citizen does not permit the teacher to teach a class in a manner that circumvents the prescribed course curriculum established by the school board when performing as a teacher … The established curriculum and LeVake's responsibility as a pubic school teacher to teach evolution in the manner prescribed by the curriculum overrides his First Amendment rights as a private citizen" (p. 502).

The Eleventh Circuit Court of Appeals has also ruled that a school can direct a teacher to refrain from discussing religion in classroom settings (*Bishop v. Aronov*, 1991), and the Supreme Court has stated that schools have a duty to make sure teachers do not inculcate religion (*Lemon v. Kurtzman*, 1971). The prohibition against an establishment of religion (as occurs with the teaching of creationism) in these situations outweighs the public school teachers' right to free speech. Finally, in *Peloza v. Capistrano Unified School District* (1994), the Ninth Circuit Court of Appeals noted that "to permit [Peloza] to discuss his religious beliefs with students during school time on school grounds would violate the

Establishment Clause of the First Amendment. Such speech would not have a secular purpose, would have the primary effect of advancing religion, and would entangle the school with religion. In sum, it would flunk ... the test articulated in *Lemon v. Kurtzman*" (p. 522).

Isn't evolution a religion? Doesn't the teaching of evolution promote the religion of evolution, and therefore violate the Establishment Clause of the US Constitution?

No, evolution is not a religion. According to the court in *Wright* (1973), the teaching of human evolution does not equal an establishment of religion by a state. This was strengthened when the Ninth Circuit Court of Appeals in *Peloza v. Capistrano Unified School District* (1994) noted that *evolution* and *evolutionism* "define a biological concept; higher life forms evolve from lower ones. The concept has nothing to do with whether or not there is a divine Creator (who did or did not plan evolution as part of a divine scheme) ... Neither the Supreme Court, nor this circuit, has ever held that evolutionism or secular humanism are 'religions' for Establishment clause purposes. Indeed, both the dictionary definition of religion and the clear weight of the case law are to the contrary. The Supreme Court has held unequivocally that while the belief in a divine creator of the universe is a religious belief, the scientific theory that higher forms of life evolved from lower forms is not (citing *Edwards v. Aguillard*, 1987) ... Since the evolutionist theory is not a religion, to require an instructor to teach this theory is not a violation of the Establishment Clause" (p. 521).

Can science teachers be required by school administrators to read aloud a disclaimer saying that their teaching of evolution is not meant to dissuade students from accepting the biblical version of creation?

In *Freiler v. Tangipahoa Parish Board of Education* (1999), the Fifth Circuit Court of Appeals ruled that it is unlawful to require teachers to read aloud disclaimers saying that the biblical version of creation is the only concept from which students were not to be dissuaded. Such a disclaimer, according to the court, was "intended to protect and maintain a particular religious viewpoint, namely belief in the Biblical version of creation." Such disclaimers are "contrary to an intent to encourage critical thinking" (p. 345) and do not advance free thinking, or sensitivity to, and tolerance of, diverse beliefs.

Some schools are inserting into their textbooks "disclaimer" stickers stating that evolution is "only a theory" or that it is "a theory, not a fact." Have the courts said anything about these stickers?

Yes. In 2005, in *Selman et al. v. Cobb County School District*, the US District Court for the Northern District of Georgia ruled that it is unconstitutional to paste stickers claiming that, among other things, "evolution is a theory, not a fact," into science textbooks. Such stickers convey "a message of endorsement of religion" and "aid the belief of Christian fundamentalists and creationists."

However, in 2006 this decision was vacated by the 11th Circuit Court of Appeals, which remanded the case for further evidential proceedings.

Many students, parents, and antiscience activists want schools to include "intelligent design" in their science curricula. Have the courts said anything about "intelligent design"?

Yes. Late in 2005, in *Kitzmiller et al. v. Dover Area School District*, the US District Court for the Middle District of Pennsylvania ruled that (1) "the overwhelming evidence ... established that intelligent design (ID) is a religious view, a mere relabeling of creationism, and not a scientific theory," and, instead, is nothing more than creationism in disguise, (2) the advocates of ID wanted to "change the ground rules of science to make room for religion," and (3) "ID is not supported by any peer-reviewed research, data, or publications." The judge also noted the "breathtaking inanity" of the Dover School Board's ID policy and the board's "striking ignorance" of ID, and made the following point: "It is ironic that several of [the members of the School Board], who so staunchly and proudly touted their religious convictions in public, would time and again lie to cover their tracks and disguise the real purpose behind the ID Policy."

Future Prospects

Creationists have lost every challenge involving the teaching of evolution and creationism in public schools. This poor record has prompted most creationists to abandon the courts as a way of inserting their religious views into the science classrooms of public schools. Instead, creationists have used a new, two-pronged approach to promote their religious and political agenda: (1) trying to include antievolution policies in state educational guidelines, and (2) despite the *Kitzmiller* ruling (see above), repackaging creationism as "intelligent design" (ID) and declaring it to be a legitimate scientific alternative to evolution.

Politics and states' educational guidelines. Creationists' newest strategies have targeted politicians and state educational guidelines. For example, in 1999 the Kansas Board of Education—an elected group that was dominated by creationists—eliminated virtually all mention of evolution in the state's science standards; soon thereafter, the Kentucky Department of Education deleted the word *evolution* from its educational guidelines. Although most of the creationists on the Kansas Board of Education lost their ensuing election and evolution reappeared in the state's educational guidelines, creationists later reclaimed the majority of the Board and in 2005 again passed science education standards that require students to study alleged "doubts" about the theory of evolution.

Today, the standards for teaching evolution in most states continue to range from "weak" to "useless," "disgraceful," "reprehensible," and "an embarrassing display of ignorance" (Gross *et al.*, 2005; Lerner, 2000). Nationwide, 20–30% of biology teachers avoid or only briefly mention evolution, and 10–15% teach creationism (Moore, 2002b), despite the fact that doing so is unlawful (see above). In the past few years, creationists have attempted to have their views inserted into science education guidelines in Ohio, New Mexico,

Georgia, Minnesota, Texas, and elsewhere (Staver, 2003). In Louisiana, the Committee for Scientific Standards lists evolution, incest, witchcraft, drug use, and the occult as topics that should be avoided on the state's exit exam for high school students (Moore, 1999). A similar situation exists in Kentucky, where evolution is grouped with topics such as gun control as topics that "may not be suitable for assessment items" (Moore, 2001c).

Although courts have struck down all attempts to introduce creationism into science classes of public schools, most politicians continue to endorse creationism. For example, virtually all of the major candidates in the 2000 presidential election endorsed the teaching of Biblical creationism, as have many previous presidents.[12] The Republican Party platforms in many states endorse creationism (Paterson & Rossow, 1999), and politicians often use pro-creationism claims to energize their supporters. For example, in 1999 US House of Representatives majority whip Tom DeLay linked the teaching of evolution with school violence, and a state legislator in Louisiana introduced a bill blaming evolution for racism (in fact, both evolution and creationism have often been used to justify racism; see Moore, 2001a, and references therein). Former presidential candidate Pat Robertson claims that scientists are involved in a vast conspiracy to hide "proof" of creationism, and others have alleged that the acclaimed PBS *Evolution* series "has much in common" with terrorist attacks in the US (Benen, 2002).

"Intelligent design" as the new form of creationism. Intelligent design (ID) is an old, well-worn theological argument (Paley, 1831; Peterson, 2002) that claims the complexity of natural systems is proof for the existence of an intelligence that has designed each species. Unlike the so-called "young Earth creationists" who claim that Earth is only about 6000 years old and that dinosaurs were on Noah's ark during a worldwide flood, ID advocates sometimes have advanced degrees in scientific fields and use academic language, thereby giving their ideas the veneer of respectability. As Pennock has noted, leaders of the ID movement are often "more knowledgeable, more articulate, and far more savvy" than traditional creationists who base their beliefs on a literal reading of the Book of Genesis.

Advocates of ID such as Michael Behe (1996) and Phillip Johnson (1997)—who Pennock (1999) refers to as "the most influential new creationist and unofficial general of ID"—are often articulate and accept some of the tenets of evolution by natural selection (e.g., age of the earth). These creationists try to undermine Darwinism by ignoring data and claiming that there is a controversy about the scientific validity of Darwinism. For example, Jonathan Wells of the antievolution Discovery Institute claims that, "There is a growing

---

[12] Years ago, presidents of the US were proud to acknowledge their acceptance of evolution and their rejection of fundamentalism. For example, Woodrow Wilson stated that "no intelligent person at this late date denied evolution," and Theodore Roosevelt (Wilson's successor) claimed to have studied natural history "at the feet of Darwin and Huxley" (Moore, 2002c).

controversy over how evidence for evolution is presented, and students should know that." Although this rhetorical appeal for fairness and open-mindedness is appealing to many people, it is flawed because *there is no scientific controversy about Darwinism* (Staver, 2003). Fairness does not require that biologists must commit educational malpractice by teaching students about religious ideas and nonexistent controversy. As biologist Kenneth Miller (1999) has noted, "Being open-minded does not mean that adding two plus two equals five should be taught in math class." Nevertheless, ID advocates continue to presume that any undermining of Darwinism will ensure that ID will be the only available alternative, thereby ignoring all of the other creation myths (Martin & Martin, 2003). This is a classic example of a logical fallacy that philosophers refer to as an appeal to ignorance (e.g., "When the premises of an argument state that nothing has been proved one way or the other about something, and the conclusion then makes a definite assertion about that same thing, the argument commits an appeal to ignorance"; Hurley, 1997). Similarly, the claim by creationists that ID is the only alternative to evolution involves the logical fallacy of the false dichotomy, which occurs "when one premise of an argument ... presents two alternatives ... as if no third alternative were possible" (Hurley, 1997).

ID advocates claim that ID is not a religious idea because the alleged "intelligence" could be anything—gods, extraterrestrials, ghosts, demons, or virtually anything else imaginable. In reality, however, ID advocates speak of their Christian god, not little green people from outer space. For example, ID leader William Dembski defines ID as a movement "that challenges Darwinism ... and [is] a way of understanding divine action" (Dembski, 1999), and Phillip Johnson admits that ID is meant to introduce people to the truth of the Bible (Benen, 2002). Although the vagueness of ID enables believers to reconcile just about anything as the force responsible for the "design" of life, it nevertheless has numerous drawbacks. For example:

- ID is not science (Staver, 2003). Advocates of ID have published no scholarly articles in respected scientific journals, and all of the science-related claims of ID have been painstakingly refuted (e.g., Miller, 1999). Advocates of ID have offered no scientific hypotheses for testing and have yet to establish their ideas as having a scientific basis (Staver, 2003). ID leaves scientific problems (e.g., global warming, the creation of new antibiotics and vaccines) completely unsolved.
- ID immediately leads to scientific dead-ends. For example, after claiming that some unidentified "intelligence" is responsible for the design of life, what is science to say? What is the intelligence? When did the intelligence design life? How did the intelligence design life? Does the intelligence continue to intervene and design life? How do we know?

Moreover, it is difficult to reconcile the inefficiencies of nature with any "intelligence." For example:

- If an "intelligence" designed all of life, why have more than 99% of all species gone extinct? Stated another way, why has virtually everything that the "intelligence" designed been such a failure? Extinctions are to be expected in Darwin's world, but they are hard to reconcile in an ID world—unless, of course, the "intelligence" is not very intelligent.

- If organisms were designed by an "intelligence," why do they have so many similarities (e.g., bone structure, mechanisms of inheritance)? Why are natural processes so wasteful? Why do organisms have structures that are seemingly useless (e.g., why do whales and pythons have pelvic bones)? Why are there so many serious flaws in nature? These vestigial structures are easily explained in Darwin's world (e.g., Darwin called them "parts in this strange condition, bearing the stamp of inutility"), but they are hard to explain with a beneficent intelligence in an ID world. As the late Stephen Gould (1980) noted, "Odd arrangements and funny solutions are the proof of evolution—paths that a sensible God would never tread but that a natural process, constrained by history, follows perforce."

Despite these and other problems, ID and other forms of creationism remain popular among the public, as well as with a surprising number of science teachers (Gallup & Newport, 1991; Moore, 1999, 2002b; National Science Board, 1996). The public's rejection of evolution has caused many science museums in the US to avoid exhibits of human evolution (Marks, 1998), and pro-science governmental agencies such as the National Science Foundation often edit their public reports so they do not include the word *evolution* (Pigliucci, 1998). Evidence supporting evolution continues to accumulate from a variety of scientific disciplines, ranging from molecular biology to archaeology (Gregg, Janssen, & Bhattacharjee, 2003), yet creationists continue to attack evolution. Just as Bryan claimed that the evolution/creationism debate would be a "duel to the death" (see above), today's creationists "see themselves as participants in a holy war against forces that would undermine the foundations of true Christianity and they see 'evolutionism' as the godless philosophy that unites the enemy" (Pennock, 1999). As Ken Ham, the director of the well-funded antievolution organization group has claimed, "There is a war going on in society—a very real battle … it's really creation versus evolution."

References

Abington School District v. Schempp, 374 US 203 (1963).
Behe, M. (1996). *Darwin's black box*. New York: The Free Press.
Benen, S. (2002, May). Insidious design. *Church & State*, pp. 8–13.
Bishop v. Aronov, 926 F. 2d 1066 (1991).
Crowley v. Smithsonian Institution, 636 F.2d 738 (D.C. Cir. 1980).
Daniel v. Waters, 515 F. 2d 485 (6th Cir. 1975).
Dembski, W. (1999). *Intelligent design: The bridge between science and theology*. Downers Grove, IL: InterVarsity Press.
Edwards v. Aguillard, 482 U.S. 578 (1987).
Epperson v. Arkansas, 393 U.S. 97 (1968).
Freiler v. Tangipahoa Parish Board of Education, 185 F.3d 337 (5th Cir. 1999), *cert. denied*, 530 U.S. 1251 (2000).
Gallup, G. H., Jr., & Newport, F. (1991). Belief in paranormal phenomena among adult Americans. *Skeptical Inquirer, 2*, 137–147.
Gould, S. J. (1980). *The panda's thumb*. New York: W. W. Norton.
Gregg, T. G., Janssen, G. R., & Bhattacharjee, J. K. (2003). A teaching guide to evolution. *The Science Teacher, 70* (8), 24–31.

Gross, P., Goodenough, U., Lerner, L., Haack, S., Schwartz, M., Schwartz, R., & Finn, C., Jr. (2005). *The state of state science education standards 2005*. Washington, DC: Thomas B. Fordham Foundation.
Hellend v. South Bend Community School Corporation, 93 F.3d 327 (7th Cir. 1996), *cert. denied*, 519 U.S. 1092 (1997).
Hendren v. Campbell, Superior Court No. 5, Marion County, Indiana, April 14, 1977.
Hurley, P.J. (1997). *A concise introduction to logic* (6th ed.) Belmont, CA: Wadsworth.
Johnson, P. (1997). *Defeating Darwinism by opening minds*. Downers Grove, IL: InterVarsity Press.
Lemon v. Kurtzman, 403 US 602 (1971).
LeVake v. Independent School District #656, 625 N.W.2d 502 (MN Ct of Appeal 2000), *cert. denied*, 534 U.S. 1081 (2002).
Lerner, L. S. (2000). *Good science, bad science: Teaching evolution in the states*. Washington, DC: Thomas B. Fordham Foundation.
Martin, B. & Martin, F. (2003). Neither intelligent nor designed. *Skeptical Inquirer, 27* (6), 45–49.
McLean v. Arkansas Board of Education, 529 F. Supp. 1255, (E.D. Ark. 1982).
Miller, K. (1999). *Finding Darwin's god*. New York: HarperCollins Publishers.
Moeller v. Schrenko, 554 S.E.2d 198 (GA Ct. of Appeal 2001).
Marks, J. (1998, December 4). How can we interject human evolution into more museums? *The Chronicle of Higher Education*, p. B9.
Moore, R. (1999). The courage and convictions of Don Aguillard. *The American Biology Teacher, 61*, 166–174.
Moore, R. (2001a). Racism, creationism, and the Confederate flag. *The Negro Educational Review, 52*, 19–28.
Moore, R. (2001b). The lingering impact of the Scopes trial on high school biology textbooks. *BioScience, 51*, 791–797.
Moore, R. (2001c). The revival of creationism in the United States. *Journal of Biological Education, 35, 17*–21.
Moore, R. (2002a). *Evolution in the courtroom: A reference guide*. Santa Barbara, CA: ABC-CLIO.
Moore, R. (2002b). Teaching evolution: Do state standards matter? *BioScience, 52* (4), 378–381.
Moore, R. (2002c). The sad status of evolution education in American schools. *The Linnean, 18*, 26–34.
National Science Board. (1996). *Science and engineering indicators*. Washington, DC: U.S. Government Printing Office.
Paley, W. (1831). *Natural theology: Or, evidence of the existence and attributes of the deity, collected from the appearances of nature*. Boston, MA: Gould, Kendall, and Lincoln.
Paterson, F. R. A., & Rossow, I. F. (1999). Chained to the devil's throne: Evolution and creation science as a religio-political issue. *The American Biology Teacher, 61*, 358-364.
Peloza v. Capistrano Unified School District, 37 F.3d 517 (9th Cir. 1994).
Pennock, R. T. (1999). *Tower of Babel*. Cambridge, MA: MIT Press.
Peterson, G.R. (2002). The intelligent-design movement: Science or ideology? *Zygon, 37* (1), 7–23.
Pigliucci, M. (1998). Summer for the gods (book review). *BioScience, 48*, 406–407.
Price, G. M. (1929). Bringing home the bacon. *Bible Champion, 35*, 205.
Scopes v. State of Tennessee, 289 S.W. 363 (Tenn. 1927).
State of Tennessee v. John Thomas Scopes (1925), reprinted in *The World's Most Famous Court Trial, State of Tennessee v. John Thomas Scopes*. New York: Da Capo Press (1971).
Staver, J.R. (2003). Evolution and intelligent design. *The Science Teacher, 70* (8), 32–35.
Stone v. Graham, 449 US 39 (1980).
Wallace v. Jaffree, 472 US 38 (1985).
Webster v. New Lenox School District #122, 917 F. 2d 1004 (7th Cir. 1990).
Willoughby v. Stever, Civil Action No. 1574-72 (D.D.C. August 25, 1972), *aff'd mem.*, 504 F.2d 271 (D.C.Cir.1974), *cert. denied*, 420 U.S. 927 (1975).
Wright v. Houston Independent School District, 366 F. Supp. 1208 (S.D.Tex.1972), *aff'd*, 486 F.2d 137 (5th Cir. 1973), *cert. denied sub. nom*. Brown v. Houston Independent School District, 417 U.S. 969 (1974).

The Warfare between Darwinism and Christianity: Who Is the Attacker and What Implications Does This Have for Education?

Michael Ruse

The classic warfare account of the clash between Darwinism and Christianity occurred at the annual meeting of the British Association for the Advancement of Science, at Oxford University in 1860, the year after Charles Darwin published his great *On the Origin of Species by Means of Natural Selection*: Samuel Wilberforce, the Bishop of Oxford, announced, "I should like to ask Professor Huxley, who is sitting by me, and is about to tear me to pieces when I have sat down, as to his belief in being descended from an ape. Is it on his grandfather's or his grandmother's side that the ape ancestry comes in?" And then taking a graver tone, he asserted, in a solemn peroration, that Darwin's views were contrary to the revelation of God in the Scriptures. Professor Huxley was unwilling to respond: but he was called for, and spoke with his usual incisiveness and with some scorn: "I am here only in the interests of science," he said, "and I have not heard anything which can prejudice the case of my august client." Then after showing how unprepared the Bishop was to enter upon the discussion, he touched on the question of Creation. "You say that development drives out the Creator; but you assert that God made you: and yet you know that you yourself were originally a little piece of matter, no bigger than the end of this gold pencil-case." Lastly, as to the descent from a monkey, he said: "I should feel it no shame to have risen from such an origin; but I should feel it a shame to have sprung from one who prostituted the gifts of culture and eloquence to the service of prejudice and of falsehood." (Huxley 1900, 1, 186–187)

Wonderful stuff! Thomas Henry Huxley, the young professor at the London School of Mines, takes on the arrogant Bishop of Oxford, Samuel Wilberforce, a leading figure in the Church of England, and comes away triumphant. The Bishop is left crushed. This time, at the legendary meeting of the British Association for the Advancement of Science in Oxford in 1860, science triumphed over religion: revenge for the time when the aged Galileo was made to recant the Copernican revolution.

But was there really this kind of warfare between science and religion? Did the theory of evolution, as announced by Charles Darwin in his work *On the Origin of Species* in 1859, really so upset the Christian Church? What are the implications of all of this for today, specifically with respect to the clash between evolution and Creationism? How should all of this affect the teaching of young people? Let us start with a bit of history and then move up to the present.

Before Darwin

Evolution is the child of progress. The eighteenth century, the Age of the Enlightenment, saw thinkers and activists in all of the major European countries—especially England, France, and Germany—increasingly enthused by the hope that through human effort it would be possible to improve our understanding of the world around us, and thus to better the social and cultural conditions of all of humanity. By one person after another, it was claimed that things are getting better, that humans are getting more knowledgeable, and that through continued effort and application there can only be an upward rise. Intoxicated by this ideology, a small but increasing number of thinkers turned their attention to the organic world. They argued that just as socially and intellectually one sees an upward progressive rise, so also this is reflected in the worlds of animals and plants: there is an ever-developing chain from the simplest form, the "monad," right up to the most complex and sophisticated form, naturally enough "man".

Paradigmatic of such thinkers was Erasmus Darwin, grandfather of Charles and in his own right an eminent British physician and friend of industrialists. He believed passionately that the cultural world was getting better, and that this is a fact to be found also amongst animals and plants. Then, in good circular fashion, he concluded that his ideology of progress was supported by his biology of evolution. Likewise, in France, the most important of all the pre-(Charles)-Darwinian evolutionists, Jean Baptiste de Lamarck, author of *Philosophie zoologique* (1809), was a friend of the *philosophes*. He believed that upward progress is possible and that this is something to be found likewise in the organic world. The history of life takes us all the way from spontaneously generated worms, through the orangutan, and finally to the human.

This evolutionism was a doctrine upsetting to many. With talk of evolution, one was certainly going against traditional Christian beliefs about origins. Does this mean that evolution was the theory of atheists? Not at all! Erasmus Darwin, Lamarck, and other evolutionists were, in their way, as religious as any. They had broken from conventional religion. In particular, they had broken from the "theistic" idea of God as a beneficent creator: a Being who hovers constantly over His creation, prepared to intervene miraculously as necessary. Rather, evolutionists (like many others) were moving to a "deistic" conception of God: they believed in a God who had created His world and who then withdrew from active work; a God who let everything unfurl, automatically as it were, like clockwork driven through the power of unbroken laws. For the evolutionists, this picture of God as unmoved mover led naturally to the idea of developmentalism. For evolutionists, indeed, their take on origins was—far from being a refutation of a Creator—the greatest evidence in His favor. Evolutionism, therefore, was not particularly Christian. Equally, it was anything but atheistic.

Charles Darwin

Charles Robert Darwin was born in England in 1809, living right through the century until 1882. As a young man, he was a sincere Christian, intended even for a career as a parson (priest) in the Church of England. However, after education at the University of Cambridge (1828–1831), Darwin went off as naturalist on board *H.M.S. Beagle*. Thus, he spent some five years away from home: mainly in South America, but eventually circumnavigating the globe. During this voyage, Darwin's religious ideas began to change. This was due particularly to the influence of the writings of the Scottish geologist Charles Lyell, whose *Principles of Geology* Darwin took with him on the *Beagle*. The Lyellian philosophy of geology, "uniformitarianism," stressed above all that God works through law rather than through miracle, especially rather than through catastrophic miracle.

Darwin took this on board, and his theistic Christianity changed gradually to a form of deism. No longer did Darwin look for evidence of God's intervention in the world. Rather, Darwin took signs of unbroken law—law creating everything in a gradual fashion—to be the greatest mark of God's power and goodness.

In 1835, the *Beagle* visited the Galápagos Archipelago in the Pacific. It was this event that turned Darwin towards evolutionism, although the actual moment of change probably did not occur until Darwin was back in England at the beginning of 1837. He was amazed that, on islands only 10 kilometers apart, one could find different species of birds and reptiles (notably giant tortoises). Darwin reasoned that the only way that this could possibly have come about, consonant with his belief in a God of law, was through the natural process of evolution. All other options imply impossible coincidences. In fact, therefore, one can say that Darwin became an evolutionist because of his religious beliefs rather than in spite of them.

Darwin knew full well that the best science not only describes but explains: it offers causal explanations. Intending to be the Newton of biology, Darwin searched hard for some 18 months for a possible causal force behind evolution. Eventually, at the end of 1838, Darwin hit upon the mechanism for which today he is justly famous: natural selection. Darwin saw that there is ongoing population growth, combined with limited space and food resources. This leads to a struggle for existence. From this, Darwin argued that there will be a constant winnowing or selecting of certain organisms: these distinctive organisms being the progenitors of all future members of the group. Eventually, all this results in a gradual form of evolutionary change.

In at least two basic respects, Darwin's Christian background and training had a crucial impact on his thinking at this point. First, while an undergraduate, Darwin had read the works of (Anglican) Archdeacon William Paley: the greatest exponent in recent times of the argument from design. Noting that the eye is much like a telescope, Paley had argued that just as telescopes have telescope makers, so eyes must have eye makers: the Great Optician in the Sky. Darwin agreed entirely with Paley's premise that eyes and other organic characteristics—known today as adaptations—seem as if they were designed. Such characteristics function for the good of their possessors. Any mechanism that Darwin endorsed was going to have to recognize this fact. Indeed, having realized that animal and plant breeders get their effects by selecting, and that a natural equivalent could do the same thing, Darwin congratulated himself on his success. Selection produces precisely the teleologically directed adaptations that Darwin and the natural theologians thought so important.

Then second, still searching for a force behind a natural form of selection, Darwin read a very conservative political doctrine by the Reverend Thomas Robert Malthus, another clergyman in the Church of England. Malthus argued that state welfare systems are not only wrong, but foolish: if you feed the poor, you only have more of them in the next generation. Inevitably, there will be a bloody clash and struggle for existence. Thus, it is better not to exacerbate the situation through misguided attempts at public charity. Darwin seized on this, generalizing from humans to animals and plants, and realizing that in nature there would be an ongoing struggle for existence. Then, turning Malthus on his head as it were, Darwin argued that this would be the force behind a natural form of the artificial selection that he had seen breeders using so effectively. Malthus' belief that God's Providence ensures that human attempts at social engineering are doomed to failure was converted into an evolutionary picture of never-ending upward progressive change.

Here, then, Darwin had his mechanism for evolutionary transformation. As the route to the fact of evolution was paved with the stones of deism, so the route to the mechanism of selection was paved with the stones of theism. *On the Origin of Species* was published some twenty years after the discovery of the mechanism. By this time, Darwin's religious beliefs were starting to fade somewhat. Indeed, it seems clear that, in the last ten to twenty years of his life, Darwin moved from deism all the way to agnosticism. He never became an atheist; but, torn by the problem of evil and pain, Darwin found that he simply could not reconcile his emotions and thinking with a belief in any kind of deity. However, in the writing of the *Origin*, Darwin certainly gave no evidence or premonition of his gradual loss of faith. In fact, he made it clear quite genuinely that, although the theory of evolution through natural selection was being presented as a challenge to conventional conservative Christianity, this was not done in an irreligious or anti-God spirit. Again and again, Darwin referred favorably to the actions of the Creator. Some say that Darwin was simply larding his science with such phrases to quell opposition. I would rather argue that these sentiments were (at the time of writing) genuinely felt. Darwin accepted a God: a God who worked through unbroken law, through natural selection bringing about change in the direction of adaptation and design-like effects.

We see that Darwin's theory was far from being a text intended simply as a manifesto in the battle between science and religion. What then happened that it should be thus taken and so quickly? How could it be that Thomas Henry Huxley, who labeled himself "Darwin's bulldog," took on the Bishop of Oxford in a battle between secular evolution and sacred Christianity? Let us see.

After Darwin

After the *Origin*, evolution was accepted almost overnight. Christians and others moved together to a transformist view of organic origins. Darwin's arguments for the fact of evolution simply overwhelmed. It is true that many felt far less comfortable about the mechanism of natural selection. They thus supplemented their evolutionism with various auxiliary mechanisms, including guided interventions by the Creator. This was a move taken, for instance, by Darwin's great champion in North America, the Harvard botanist Asa Gray. Indeed, almost all religious people wanted to say that humans require something special: miraculous creations of souls, in particular. However, evolution as such became part of common intellectual currency. Moreover, had this been all there was to it—especially given that human souls are hardly part of science anyway—then quite probably no one today would think of the Darwinian Revolution as, in any sense, being something intimately involved in the battle between science and religion.

However, there was another important factor at play. Darwin's scientific supporters—Huxley and his friends—were fighting hard in the second half of the nineteenth century to reform British society. They wanted a meritocracy: a culture where people could succeed by talent and effort; where education would be freed from religious conditions; where the civil service would be well trained and promotion would be on performance; where the army would no longer be ruled simply by aristocrats without any sense of feeling for those beneath them; where all of these things and more would be made possible. In short, Huxley and his friends wanted to turn the world from the feudal eighteenth century into the modern industrialized twentieth century.

In order to do this, they saw that positive moves would not be enough. They would also have to fight the reactionary forces of the past. Most notably, they saw the Church of England (the Anglican Church) as representing one of the greatest barriers that they had to overcome. They saw the Church as intellectually moribund. Socially, they regarded it as a dinosaur. Unfortunately, as it may have been with dinosaurs, it was dangerous if alarmed and attacked! The reformers therefore were looking for an alternative ideology—an alternative secular religion, if you will—with which they could promote and justify their own actions. Evolution was the perfect vehicle. As we have seen, it was impregnated with ideology from its very beginnings: the ideology of progress.

This was taken up, modernized, supplemented with other ideals, and then presented to the world as a new way of seeing things: a scientific way of seeing things, a modern way of seeing things, a throw-away-Christianity-and-all-the-other-outmoded-ways of seeing things. Even though Darwin himself had certainly not intended his theory this way, this was the way that his supporters set out to use his ideas.

Thus, in the years after the *Origin*, we find that evolution was hijacked by people determined to use it in opposition to conventional Christian religion; people determined to use it less as a scientific theory and more as a secular ideology. Hijacking Darwin's name also, this "Social Darwinism"—generally demanding *laissez-faire* individualism—was promoted as the panacea for all of society's ills. One could almost have forecast, therefore, that this would raise the ire of many Christians, particularly those of a conservative bent. This, of course, is precisely what happened, particularly in the US. In the US, at the beginning of the twentieth century, there developed a particularly virulent form of Protestant Christianity—fundamentalism—which claimed that the true essence of Christianity is to be found in and only in the Bible taken absolutely literally. Hence, the scene was well and truly set for a massive science–religion conflict: Darwin versus God. Something of an irony, as you will surely now realize, for even if God wanted a good fight, Darwin would have wanted no part of the battle. It was something he was forced into by others who used his name and ideas for their own ends.

Evolution and Religion Today

It was around 1930 that a number of very professional scientists took over the leadership of evolutionary thought. They were determined to move it from its status as a secular ideology and into a new role as a fully professional, university-based science. Therefore, thanks particularly to the labors of such people as Sir Ronald Fisher in England and the Russian-born Theodosius Dobzhansky in the US, evolution finally began to take its place as a fully functioning branch of biology, with the kinds of standards that one expects elsewhere in science, as in physics and chemistry. With this move towards professionalism, and away from secular ideology, we find that evolutionists themselves started to drop many of their religious and ethical exhortations. Even if they were progressionists, evolutionists took major care to keep such sentiments out of their strictly scientific writings. Naturally, with this change, evolution became a lot less threatening as an ideology, and the conflict between it and Christianity subsided significantly.

One can say that truly, in the minds of many, both Darwinians and Christians, it is possible now to have a happy synthesis between science and religion; a synthesis, incidentally, endorsed by John Paul II, who has recently written favorably about evolution—even its mechanisms—from a Christian perspective. Unfortunately, however, some do live very much in the past, still fighting the battles of Thomas Henry Huxley and friends. The prime example is Richard

Dawkins, Oxford biologist and very popular science writer. Not only does he write vigorously in favor of a pure Darwinism, but again and again he stresses that such Darwinism makes Christian belief positively untenable. He speaks of those who think you can have both science and religion as infected with "intellectual flabbiness." For Dawkins, Darwinism emphasizes the struggle for existence and the vile nature of so many adaptations: adaptations, for example, of the parasite living on its prey. In no way can one reconcile this with a god of love and care—the Christian God. Hence, one can accept it as Darwinism or Christianity, but not both. In Dawkins' opinion, the fundamentalists may have been wrong, but they alone were honest in seeing that you must make a choice between two masters. "Given a choice between honest to goodness fundamentalism on the one hand, and the obscurantist, disingenuous doublethink of the Roman Catholic Church on the other, I know which I prefer" (Dawkins, 1997a, p. 399). In fact, having himself characterized his move to atheism from religious belief as a "road to Damascus" experience (Dawkins, 1997b), Dawkins is ecumenical in his hostility towards Christianity. Explicitly, he uses terms that we often associate with the hatred of one religion by subscribers to another: "I am considered by some to be a zealot. This comes partly from a passionate revulsion against fatuous religious prejudices, which I think lead to evil. As far as being a scientist is concerned, my zealotry comes from a deep concern for the truth. I'm extremely hostile towards any sort of obscurantism, pretension. If I think somebody's a fake, if somebody isn't genuinely concerned about what actually is true but is instead doing something for some other motive, if somebody is trying to appear like an intellectual, or trying to appear more profound than he is, or more mysterious than he is, I'm very hostile to that. There's a certain amount of that in religion. The universe is a difficult enough place to understand already without introducing additional mystical mysteriousness that's not actually there. Another point is esthetic: the universe is genuinely mysterious, grand, beautiful, awe inspiring. The kinds of views of the universe which religious people have traditionally embraced have been puny, pathetic, and measly in comparison to the way the universe actually is. The universe presented by organized religions is a poky little medieval universe, and extremely limited." "I'm a Darwinist because I believe the only alternatives are Lamarckism or God, neither of which does the job as an explanatory principle. Life in the universe is either Darwinian or something else not yet thought of." (Brockman, 1995, pp. 1985-1986)

How to Respond?

The question I want to ask now—and this is the key question of this chapter and the reason for writing it—is what kind of response should one make to stuff like this? Let me say that today's Creationists—both the traditional kind, who believe in six days of creation and so forth, and the more modern versions including the so-called Intelligent Design (ID) theorists—are delighted when Dawkins and his fellows emote like this. The Creationists know that Christians of every kind find Dawkins deeply offensive and hurtful—as, of course, he intends. They also know that Christians of every kind know full well that Dawkins is an ardent Darwinian, in addition to being an ardent atheist. Finally, they also know that Christians are aware that Dawkins, as in the passage above, links his Darwinism and his atheism. Hence, the Creationists know—to their great delight as I have just said—that every time Dawkins opens his mouth, he is setting up prejudice against Darwinism in almost every stripe of Christian.

Hence, either people want to reject Darwinism, or—and this is especially true in the US—they want to go all the way and reject evolution, or remain secure in their having already made the right decision in rejecting evolution. Dawkins has been a wonderful champion of Darwinism. Unfortunately, when he emotes like this, he is doing nothing but harm—especially in the US today—to the cause of all kinds of evolution, and certainly making much more difficult the cause of those of us who are fighting to keep Creationism (including ID) out of the schools. All that the ID people have to do is show passages from Dawkins to their local school board or their congressman, and all hell breaks out. This is a fact.

Now, obviously, if you are someone who genuinely distrusts Darwinism—someone who, for instance, believes in self-organization rather than natural selection—one strategy is simply to distinguish one's own kind of evolution from that of Darwinians, including that of Dawkins. This is a move taken by several of my good Christian friends, including Holmes Rolston III (1987, 1999) and David Depew and Bruce Weber (1994). They are not cowards, so I do not think they turn from Darwinism because they are scared of Richard Dawkins. But I do know that they are pleased and relieved that they can sidestep Dawkins and company, and can get on with the work that interests them. In good conscience, they can stand up to school boards and congressmen and make the case for evolution as they understand it.

This is also a move that can be taken by non-Darwinians who are not Christians. The late Stephen Jay Gould had an extremely ambiguous relationship with Darwinism—he thought there was some merit in it, but by no means was he overawed by its potential (Gould, 1977, 1989, 2002). He preferred to stress the non-Darwinian nature of the living world as much as its Darwinian aspects. He too could stand up to school boards and congressmen and make the case for evolution as he understood it. No one could think that he was linking Darwinism and atheism—or even that he was linking evolution and atheism (Gould, 1999). He took care to argue that science and religion are different fields of inquiry (different "Magisteria," as he called them). Although I am not sure that he was always as good at keeping them separate as he claimed, at least he had a strategy—a strategy incidentally shared by many Christians, especially those who are inclined to a version of Karl Barth's neoorthodox theology.

What if you are indeed as enthused by Darwinism as is Richard Dawkins? Well, if you are a Christian, then you can argue right back and deny that there are the links that Dawkins forges between Darwinism and atheism. This is the position taken by someone like the Cambridge paleontologist Simon Conway Morris (2003). He is a serious scientist whose work on the Cambrian fossils of the Burgess Shale is acknowledged to be absolutely top quality. He is also a practicing Anglican and more than willing to come out fighting for his faith. His most recent book, *Life's Solution*, argues that humans are probably a unique creation but that our special attributes including (especially) intelligence are just the sorts of things one would expect from the workings of blind natural law. He believes that a Darwinian approach to nature can give traditional Christians all they want and more. So again, we have someone who can stand up to (the English equivalents of) school boards and congressmen and argue that Darwinism in no way is going to threaten people's religions beliefs.

What if you are a Darwinian and a nonbeliever? What if you have as little religious belief as Dawkins? I fall into this category. I grew up as a Christian—a Quaker, actually—but from about the age of twenty, my religious beliefs have been minimal or nonexistent. I think at one point I would have described myself as a militant atheist like Dawkins. Now I am probably best described as an agnostic, although I prefer the term skeptic. So often, agnostics are people who are bored or indifferent to religion (like my wife), whereas I am very interested and my skepticism is a product of much thought rather than little or none. I am certainly an atheist with respect to most of Christianity—virgin birth, miracles, resurrection, the Trinity, and so forth.

There are two possibilities here. On the one hand, you might argue that whether or not you yourself are a believer, you think that a Darwinian can be a Christian. This is my position, and I have written a book on just this very topic. Of course you cannot perjure yourself for political or pedagogical advantage, but if you can make this case then again you can speak to others, including school boards and politicians. You can argue—as I have argued many times, privately and publicly—that Darwinism is a scientific theory and that you can hold it whether you have Christian belief or not. On the other hand, of course, you might go all the way with Dawkins and argue that Darwinism implies atheism, that you accept the former (Darwinism), and that therefore in good conscience you have to accept the latter (atheism)`.

What Should Richard Dawkins Do?

What then? Someone might argue that, out of prudence, Dawkins ought to keep his beliefs to himself. Even if this were possible, I must confess that I myself feel very uncomfortable with this advice. If Richard Dawkins or someone like him feels that they can link Darwinism and atheism in the way that they can, and if they feel as does Dawkins that religious belief is a positive evil—leading to 9/11 and the like—then I would be the last to tell them to shut up. I will just mop up as best I can, after they are done; perhaps before they are done. I have to be polite to them, I have to be tolerant of them, but I do not have to agree with them or stay silent.

However, I do think we have the right to make some demands of people like Richard Dawkins, and I think they have the obligation (given that their actions are so harmful to the cause of teaching evolution) to consider our demands seriously. First, they should think about the history of evolutionary theory—the history as I have given it in the first part of this paper. They should ask themselves if there is still need (as Thomas Henry Huxley thought there was) to continue the warfare between science and religion. The battles Huxley was fighting—for universal education, for proper medical training, for a qualified civil service—are won. Even if one agrees that there was at one point a need for a clash between science and religion, it is worth asking if there is still such a need. People like Richard Dawkins should ask themselves if they are caught in a time warp, fighting the wars of the nineteenth century. If they came to realize that they are, then perhaps they themselves might want to modify their behavior—their militancy, at least.

Second, such people should think about the contemporary state of religion. I am always amazed at how ignorant Dawkins (and others such as the philosopher Dan Dennett and the historian Will Provine) are of modern Christianity—or indeed, of the history of Christianity. Take the question of religion and violence. I agree that 9/11 was in major part a function of vile and totally ridiculous religions convictions—although anyone who thinks that the hatred of the US today is simply a function of religion is naive to the point of (as Dawkins would say) wickedness. To make the counter case, you could start with Israel, move on to Iraq, back to Chile, and then keep going.

I am not saying that the US is the new evil empire—I live by choice in the US, and I know its great virtues. Much as I love the land of my birth, I would rather live on this side of the Atlantic. I am simply saying that there is more to these issues than religion—pure and simple. Anyone who thinks that religion, misguided or not, has never led to good things is again naive to (in my opinion, beyond) the point of wickedness. I could talk of Elizabeth Fry, the nineteenth-century Quaker, who did so much for prison reform—because she was a Christian. I could talk of the White Rose group in Nazi Germany whose lives were ended by the knife of the guillotine, because they spoke out for their Christian beliefs—or of Dietrich Bonhoeffer whose life ended on the scaffold because of his insistent determination to follow the gospels. I could talk of those brave Christians and Jews who stood against apartheid in South Africa. I could talk today of the Anglicans and others in downtown Tallahassee who run the shelters and kitchens for the alcoholics and the handicapped and the hopeless of the city. They do it because they think that that is what their Savior demands of them. I do not think Jesus was their savior, but I am humbled by their behavior.

Turning to the realm of sociology and politics, the Darwinian anti-Christians should look much more carefully at religion, asking about its effects in the world. Mention Northern Ireland certainly. Mention South Africa also. Do not forget that the worse excesses of the twentieth century—Nazi Germany, Soviet Russia in the 1930s, and Cambodia more recently—were not done in the name of religion, but for ideologies that Christians would rightly have labeled "pagan." This leads to a second obligation at least, namely to see what is the true state of Christian (and other religious) belief, rather than what one thinks it might be, or what one remembers it to be from one's childhood memories. One may not change one's beliefs, but if one is going to condemn, then one has some duty to understand what one is condemning—especially if one's actions are going to have political consequences and pedagogical fallout for the children of the coming generations.

I would say that there is a third obligation also, namely that if—after serious study—one continues to think that Darwinism implies atheism; if one continues to think that religion is indeed an unmitigated evil and that this truth must be shouted from the rooftops; then, one must ask oneself what one is going to do about the political and pedagogical issues. How are you going to fight the threat of Creationism and ID in the schools? Or are you going to be like the Pharisee in Jesus' parable and pass by on the other side?

This is the final question I leave for my fellow Darwinian nonbelievers.

References

Brockman, J. (1995). *The third culture: Beyond the scientific revolution*. New York: Simon and Schuster.
Dawkins, R. (1997a). Obscurantism to the rescue. *Quarterly Review of Biology* 72: 397–99.
Dawkins, R. (1997b). Religion is a virus. *Mother Jones*.
Dembski, W A. (1998). *The design inference: Eliminating chance through small probabilities*. Cambridge: Cambridge University Press.
Depew, D J, and B H Weber. (1994). *Darwinism evolving*. Cambridge, Mass.: MIT Press.
Gould, S J. (1977). *Ontogeny and phylogeny*. Cambridge, Mass.: Belknap Press.
Gould, S J. (1989). *Wonderful life: the burgess shale and the nature of history*. New York: W. W. Norton Co.
Gould, S J. (1999). *Rocks of ages: science and religion in the fullness of life*. New York: Ballantine.
Gould, S J. (2002). *The structure of evolutionary theory*. Cambridge, MA: Harvard University Press.
Huxley, L. (1900). *The life and letters of Thomas Henry Huxley*. London: Macmillan.
Lamarck, J B. (1809). *Philosophie zoologique*. Paris: Dentu.
Lyell, C. (1830–1833). *Principles of geology: Being an attempt to explain the former changes in the earth's surface by reference to causes now in operation*. London: John Murray.
Morris, S.C. (2003). *Life's solution: Inevitable humans in a lonely universe*. Cambridge: Cambridge University Press
Rolston III, H. (1987). *Science and religion*. New York: Random.
Rolston III, H. (1999). *Genes, Genesis and God: Values and their origins in natural and human history*. Cambridge: Cambridge University Press.

Capturing the Educational Potential of 'Creation Science Debates'

David Mercer

Introduction

In the following discussion, I will consider the way debates about science and religion, and the creation science debate in particular; often encourage a discourse that is excessively preoccupied with attempting to resolve larger debates in epistemology involved with demarcating science and identifying and describing the philosophical essences of science and religion. Why I describe these preoccupations as "excessive" is (1) that they frequently involve trying to resolve simplistically, expeditiously, and at the expense of any contextual history of science and religion, complex and largely intractable debates in philosophy; (2) that these preoccupations also contribute to poor "public" and policy understandings of science, most notably in legal and media discourse; (3) that concerns raised by creationism have been used as a resource by some to discourage science educators from drawing on more sophisticated understandings of science developed in science studies: understandings, which have the potential to avoid some tensions flowing from creation science debates.

To support these claims I will first outline a brief critique of the idea that simple contrasts can be drawn between science and religion. Next, I will provide four brief case studies from a variety of contexts where creation science debates have been played out that illustrate how preoccupations with defining the essences of science and religion can become inconclusive and distracting. These case studies will be discussed under the following four headings: (1) attentive popular science, (2) legal commentary, (3) creation science/science and the media and (4) "anti-science studies." I will conclude by providing a preliminary sketch of some of the ways science education and popularization of science could be improved by paying more attention to science as local knowledge and practice.

"Contrasting" Science and Religion

Historian of science John Hedley Brooke has noted three typical largely inadequate contrasts that are frequently drawn between science and religion:

> Science ... operates within a "worldview" that regards natural phenomena as the product of impersonal forces. By contrast, religious and magical systems involve personalized gods, spirits, or demons. Whereas the scientific enterprise is legitimated by agreed testing procedures, the theological enterprise has been characterized by dogmatism.

> Whereas religions have required worship, ceremony, and sacrifice, these are forms of activity alien to Western Science.
>
> (Brooke, 1991, pp. 17–18)

Numerous examples can be used to encourage reflection on the adequacy of drawing such binary distinctions between the supposed essences of science and religion (Wilson, 2002).

To complicate Brooke's first contrast that science relies exclusively on naturalistic explanation, we can consider the numerous conflicts within various sciences at different times about whether or not explanations are truly naturalistic (Brooke, 1991). It has been historically shown that scientific ideas are linked in important ways to their wider social contexts. This means that even if the ideal aim of scientific practice is to only use naturalistic explanations, such explanations will still carry with them linguistic and rhetorical resonances reflecting the social context of their emergence and acceptance. For example, Darwin's appeal to the idea of the "survival of the fittest" as the causal motor for his theory of evolution by natural selection was both an attempt to provide a naturalistic explanation for evolutionary process and also a metaphor self consciously borrowed from the political writings of Herbert Spencer (Young, 1985). More recently, sociobiologists have been critiqued for introducing anthropomorphic and gendered presuppositions into their attempts to analyze human behavior (Gould, 1977).

Moving on to consider Brooke's next contrast that, unlike religious dogmatism, testing and organized skepticism are important ingredients of science, a closer empirical examination suggests that scientific practices are much more diverse (Mulkay, 1979). Studies of scientific communities both in historical and contemporary contexts have in fact suggested that dogmatism has not been an unusual feature of science (Barnes, 1982) and that the significance of testing, the types of tests possible, the prominence of fundamental untestable laws and assertions vary between different areas of science (Collins and Pinch, 1993). To complicate these stereotypes further, whereas formal doctrines of empirical testing may have not played a significant role in major religious beliefs, Brooke points out that "self criticism and renewal" have often encouraged sectarianism and reform movements and have been one of the challenges faced by many orthodox religious institutions. A number of empirically orientated studies of creation science movements (Nelkin, 1977; Locke, 1994) have also noted the similarity between more sophisticated creationists and scientists in their styles of rhetoric and appeals to ideals of scientific method. These studies suggest that dismissing all creation science claims out of hand on the basis that their aims are fundamentally dogmatic and different to science relies on a caricature of both science and creation science. It is also worth noting that even at a broader level of "ideal" images of the cognitive orientation of science, complexities still appear. As Margaret Wertheim has noted, claims that science has an ideal commitment to privilege certain types of material and mathematical explanations is in itself an article of faith; an assumption about the way the world is; a commitment that cannot itself be tested or challenged:

> The idea that the ontology of the universe can be explained by mathematical laws is itself a faith position, as is the notion that the origin of life can be explained from the laws of chemistry. These are articles of faith that I share, yet we who believe in purely material accounts of our origins must be honest that this is not a logically demonstrable position. Creationists have their beliefs, we have ours.
>
> (Wertheim, 1999, p. 22)

Considering Brooke's final contrast that science is an archetypically impersonal enterprise lacking in ritual, numerous scientists across history have spoken of scientific enterprise as a way of partaking in God's work, or seeing nature through the "mind of God" (Brooke, 1991; Nelkin, 2004). Attempts to popularize science have also led to the public projection of scientific work in ritualized terms replete with quasi-religious symbols and resonances. To support this, we need go no further than to consider the exhibits appearing in many science museums (Gregory and Miller, 1998) or revisit some of the foundational texts of enlightenment humanism such as those by Comte, which reappropriated religious ritual in a detailed and quite deliberate manner (Brooke and Cantor, 1998).

A critic could respond to these examples by suggesting that they are not indicative of the dominant "day-to-day" patterns of scientific and religious interaction that are unproblematic and function with a tacit acceptance of many of the kinds of dichotomies that Brooke challenges. But this point simply does not help us understand a number of contemporary areas of scientific controversy involving things such as human cloning, genetic modification of organisms, and evolutionary psychology, where the disputes over setting boundaries between science and society involve metaphysical as well as instrumental concerns. The tendency to draw boundaries between science and religion as if they are natural and unproblematic ignores the internal disagreements typical of areas of new science, the challenge that such science may constitute to forms of religious thought, which itself can be fluid, and disguises political commitments influencing where boundaries are set (Gieryn, 1998). To again quote Wertheim:

> If Christians have been encroaching onto scientific territory, more and more scientists are encroaching on ... issues of morality and meaning, of human behavior and motivation. Most notably, evolutionary psychologists are increasingly claiming that qualities such as altruism, filial affection and sexual preference can be explained by genetic and evolutionary means. Neuroscientists are even claiming to explain mystical experiences in neurochemical terms.
>
> (Wertheim, 1999, p. 22)

Attentive Popular Science

Debates about evolution and religion have encouraged a significant body of "popular scholarly" discussion. Much of this discussion has been aimed at a "science attentive" audience ahead of specialist academic audiences. Publicly 'visible scientists' such as the late paleontologist Stephen Jay Gould and eminent geneticist Richard Dawkins provide good examples of commentators in this genre (Gregory and Miller, 1998). One of the most dominant features of this genre has been for analysis to be prefigured by attempts to define the essential differences between the two relevant domains: science and religion. Although such an exercise offers an intuitively plausible starting point for discussion, as I have shown above, trying to arrive at essential incontestable definitions of science and religion is a deceptively difficult exercise. In a commentary on recent moves to encourage improved dialogue between science and religion, Eugenie Scott, executive director of the US *National Center for Science Education*, provides a sophisticated example of how the types of "contrasts" critiqued above permeate "attentive popular science" discussions of creation science (Scott, 2003).

Scott contends that science and religion have historically interacted in four main ways: (1) warfare, (2) separate realms, (3) accommodation, and (4) engagement. In the warfare approach, there is no middle ground, coexistence is discouraged since religious thought is interpreted as irrational and an obstacle to scientific and human progress. Scott includes classic exponents of the "conflict thesis," such as A. D. White and modern figures such as Johnson and Dawkins in the warfare camp. The most notable exponent of Scott's second "separate realms" approach is Stephen Jay Gould, who believes that religion and science serve different purposes that, if properly recognized, would remove the basis for conflict Gould has described this "proposed settlement" using the acronym *NOMA*: Non Overlapping Magisteria (Gould, 1999). Scott's third approach, the "accommodation model," suggests that theological understanding can benefit from engagement with science and theology reinterpreted in light of scientific findings. Scott notes that this has normally been the case of religion reorienting key beliefs to satisfy new scientific insights. Her fourth approach, the "engagement model," suggests a more even relationship where both domains can enrich themselves by asking new questions and combine epistemological insights. One imagines that somewhere in the accommodation and engagement models would be located many "new-age" perspectives and the theological speculations of scientists such as Paul Davies who see evidence for the existence of God in the intricacies of contemporary physics (Brooke and Cantor, 1998).

Scott's categories do capture some of the different ways science and religion have interacted, but she nevertheless undermines the value of these observations when she follows them by calling on the reader to remember that despite such religious diversity there is still "only one science"(Scott, 2003, p. 114).

The idea that there is "only one science" is a message that would be contested by a diverse cross section of historians and philosophers of science. I will return to consider this theme later. Aside from these broader historical and philosophical objections, Scott's insistence that there is only one science is also difficult to sustain in the light of more immediate considerations, including some of her own work published elsewhere. For example, in an earlier paper Scott (1997) promotes the notion that one of the potential ways that conflict between science and religion can be diffused in educational contexts is for science teaching to be "methodologically materialist" in its orientation (deny the relevance of metaphysical forms of explanation to understanding scientific domains), but avoid being "philosophically materialist" in its orientation (avoid denying the relevance of other nonscientific materialist forms of explanation for other nonscientific domains). Further, undermining her contention that there is only one science, she notes that a number of scientists approach the matter differently:

> Vocal proponents of evolutionary materialism such as William Provine at Cornell, Paul Kurtz at the State University of New York, Buffalo, and Daniel Dennet at Tufts vigorously argue that Darwinism makes religion obsolete, and encourage their colleagues to argue likewise. Although I share a similar metaphysical position, I suggest that it is unwise for several reasons to promote this view as 'the' scientific one.
> (Scott, 1997, p. 518)

Scott could have added the well-known differences between scientists such as Richard Dawkins, who has for many years vigorously proposed that science has and should take precedence over forms of religious thought (Dawkins, 2003), and the late Stephen Jay Gould, who, as noted above, argued for mutual respect for the philosophical differences between the two domains. To suggest that there is only one science glosses over the way eminent scientists such as Gould and Dawkins have "set" the boundaries between science, society, and religion in completely different ways. It is also unclear whether Scott's map is an attempt to capture the actual relations between science and religion as practices, or rather the different ways some commentators have characterized the relationship between science and religion.

Returning to consider Brooke's "contrasts," a more detailed history of the complex interactions between science and religion readily exposes how the types of dichotomies identified above, and the desire to specify essences of science (and religion), do not fit neatly with a more contextual investigation of the histories of science and religion. A more complex model also helps explain why even sophisticated attempts to map the relations between science and religion such as Scott's are difficult to fashion in any consistent and reliable way.

Legal Commentary

An informative parallel discourse to the one discussed above involves the numerous attempts to define the essence of science and its contrasts to religion in the various legal contexts that have arisen in relation to disputes about the teaching of evolution and creation science. Again, most of these disputes have taken place in the US, although there have been some notable "variations on the theme" elsewhere.

Gieryn, Bevins, and Zehr compared the images of science produced in two famous American creation science cases: the *Scopes* in the 1920s and *McLean* in the 1980s (Gieryn,Bevins, and Zehr,1985). They identified that a central theme in the professional rhetoric of science during the "Scopes era" of the 1920s was partly to show the centrality and utility of science to American culture but also to emphasize that science was still complementary to religion. By the "McLean era" of the 1980s, the professional rhetoric of science was much more exclusionary. Its central preoccupation was to exclude pseudoscience and maintain control over funding. Each form of rhetoric associated with different cultural eras represented science's differing professional "self-reflections." The preoccupations of *McLean* would still appear to be predominant (Shermer, 1997).

The *McLean vs. Arkansas* case provides a valuable microcosm to consider the problems noted in the last section. Here, historian and philosopher of science Michael Ruse acting as an expert witness helped contribute to a definition of science that was ultimately adopted by the court:

> Science (1) is guided by natural law; (2) is explanatory by reference to natural law; (3) is testable against the empirical world; (4) Its conclusions are tentative; and (5) is falsifiable.
> (Ruse, 1996a)

Ruse's "definition," aside from its functional success in the specific context of the *McLean* case, raised the concerns of some of his philosophical colleagues who suggested that he had helped develop an unsound philosophy of science simply tailored to impede creation science being accepted by the court. Larry Laudan (1982) and Michael Quinn (1984) suggested that Ruse's model of science was unrepresentative of debates within the philosophy of science, in particular its reliance on the demarcation criteria of falsification drawn from the work of Sir Karl Popper. Echoing Wertheim, Lauden noted the prevalence of unfalsifiable assertions that lie at the core of many important areas of science. He also noted that some claims by creationists made in *McLean* may have, in fact, passed the criteria of being falsifiable and thus count as being framed in scientific terms, which obscured the more salient issue, that the claims were simply empirically unsustainable: something that could be relatively easily demonstrated without the need for absolute philosophical demarcation criteria between science and non-science.

Despite the diversity of different philosophical models that can be developed to attempt to demarcate science from non-science, and the philosophical challenges involved, a number of legal commentators still use creation science debates as a vehicle to promote philosophically incoherent and syncretic legal definitions of science tailored to deny creation science scientific status in legal contexts. A recent discussion of the legal status of "intelligent design" as science or religion provides a good example of such expeditious folk epistemology. Theresa Wilson, Assistant Appellant Defender, Office of the State Defender, De Moines, Iowa (Wilson, 2003) argues that the courts need to go beyond *McLean* and adopt a more comprehensive and onerous definition of science. She proposes criteria that contain logical, empirical, sociological, and historical dimensions:

> To satisfy the logical criteria, a theory must be a 'simple, unifying idea that postulates nothing unnecessary, have a consistent internal logic, be logically falsifiable, and have clearly expressed boundary conditions. To satisfy the empirical criteria a theory must be empirically testable or lead to predictions or retrodictions that are empirically testable and verifiable, concern reproducible results, and provide criteria for the interpretation of data. To satisfy the sociological criteria, a theory must solve problems or questions unresolvable (sic) with preexisting theories, pose new problems for scientists to study, and a "paradigm" including new concepts and definitions by which problems can be studied. Finally, a theory must satisfy historical criteria by surpassing all previous theories in its explanation of relevant phenomena and is consistent with already validated ancillary scientific theories. [*Footnotes omitted from the quotation*]
>
> (Wilson 2003, p. 14).

There would appear to be little cognizance on the part of Wilson that such a definition may be philosophically incoherent, for example, linking falsifiability, empirical confirmation, and possession of a paradigm, plus a mix of so-called sociological and historical factors, means a theory has to satisfy a strict set of criteria drawn from competing and contradictory philosophies of science. Demarcation criteria "purpose built" to dismiss creation science or intelligent design may also be double-edged in their application and lead to problems in neighboring areas of law where definitions of science are also sought. For instance Popper's falsification criterion has been a favorite method doctrine of some to critique Marxism and psychiatry and some creation science claims, but falsification can also be used to cast doubts about the scientific veracity of some aspects of evolutionary biology, where arguably a number of key concepts are difficult to test in any simple way. A similar problem and cluster of issues has surrounded the attempts by the US Supreme Court to define science following the widely known *Daubert* case in 1993 (Edmond and Mercer, 1996, 1998). Interestingly, in *Daubert*, where the court's main preoccupations were with toxic torts, a similar checklist to *McLean* was proposed but without reference to natural law (Edmond and Mercer, 2002). *Daubert's* reference to Popper and testing may have been a factor leading to it becoming more difficult to have certain tort claims achieve entry to court but has had the unanticipated effect of creating admissibility problems for previously unchallenged forensic science, such as fingerprint evidence (Cole, 2001).

Producing "purpose-built" definitions of science for legal contexts simply to block creation science has obvious negative implications for the broader public and policy appraisal of science. Aside from encouraging the incoherence noted above, it does little to encourage more informed assessments of the implications of science in society more generally or the day-to-day practices of science, and merely nurtures a fragile scientism unlikely to be sustained in the face of scientific disagreement and the energetic diversity of contemporary scientific practice.

Creation Science/Science and the Media

Although not purporting to provide a comprehensive content analysis of media treatment of creation science disputes, a number of studies indicate that yet again the tendency appears for the essences of science and religion to become the currency of discussion ahead of more practical local/immediate issues. In particular, the media has displayed a tendency to portray creation science and science and religion issues by using familiar mythical tropes such as the "trial of Galileo" and the "Scopes trial" (Larson, 1997; Edmond and Mercer, 1999). For example, in analysis of the media treatment of the *McLean* trial, discussed in the last section, Marcel La Follete (1983) noted the regular framing of media reports with *Scopes* imagery.

These science myths, themselves frequently based on historical misunderstandings, are drawn upon to portray creation science disputes as part of a longstanding generic process of the battle between reason and superstition. In using such myths, the political particulars of given disputes, including the public status of science and various bodies of religious thought, are easily misrepresented. For instance, in relation to the Scopes trial, Gould (1999) has pointed out that the 1920s biology text(s) that were being promoted by Scopes supporters contained numerous references to eugenics and theories of race that would today be unacceptable and even in their own time were controversial. Edward Larson's detailed history (Larson, 1997) of the Scopes trial also reveals the intricate play of US politics and public culture that helped elevate the dispute to take on public significance well beyond questions of science vs. religion.

In the study of the media treatment of a "creation science" case, this time in Australia, Edmond and Mercer (1999) noted that *Scopes* was continually used as a mythical resource to frame media accounts. The case in question, like *McLean*, nevertheless possessed considerable differences and local legal significance, to sketch the details as briefly as possible. The case emerged from the efforts of the *Australian Skeptics* and an eminent Australian geologist to use the Australian *Federal Trade Practices Act* to restrict the capacity of a fringe religious figure to give public presentations to help raise funds for an expedition to survey what he alleged were the remains of Noah's Ark. *The Skeptics* also hoped this would become a test case to restrict the "educational" and commercial activities of organizations promoting other "fringe science" claims such as astrology and alternative health and an opportunity to reinforce the public image of the rationality of science. In practice, the trial ended up hinging on quite different more subtle issues concerning the appropriate scope of the *Federal Trade Practices Act* to cover religious activities and whether its application in matters of conflicting beliefs and ideologies would inhibit freedom of speech. Despite the way the case actually proceeded, the media persisted in treating the case as a "showdown" between science and religion and an opportunity to promote the notion that modern science was under threat from creation science. Below, I have provided three quotes where journalists drew on a mythical treatment of the Scopes trial and used it as a resource to convey polarized images of science vs. religion:

> The standing-room-only crowd expected the Monkey Trial Mark II, a 1990s version of the 1925 Scopes "Monkey Trial." They wanted to see Charles Darwin and the Bible go one on one.
> (Dayton, 1997, p. 6)

> The creationist dispute has been raging ever since the infamous 'monkey trial' in Tennessee 72 years ago, at which John Scopes was convicted for teaching evolution.
> (Pockley, 1997a, p. 529)

> Sackville recalled how the Scopes trial was extended when the prosecutor, William Jennings Bryan, took the witness stand to defend the literal truth of the Bible. ... (As loosely depicted in the 1960 Hollywood film *Inherit the Wind*, Frederick March portraying Bryan, in real life, was destroyed by defence attorney Clarence Darrow, played by Spencer Tracy, and died 2 days later.)
> (Pockley, 1997b, p. 149)

The Scopes trial continues to appear as part of the popular culture of creation science disputes. In a recent dispute over teaching biology in Kansas, a local Topeka theatre group organized the staging of the play, based on the Scopes trial *Inherit the Wind*, and the play gained considerable media attention (Biles, 2000). As Larson has noted, the play *Inherit the Wind* took considerable license with the facts of the case so as to project as powerfully as possible the irrationality of opponents to evolution; the play yet again had its own "cold war" political context (Larson, 1997).

These brief examples obviously do not constitute a comprehensive overview of the media treatment of creation science but they fit in well with some of the broader critiques of the tendency for the mass media to anchor its treatment of science in simplistic narratives that fail to provide a sense of the social context of science (Gregory and Miller, 1998). These examples are also consistent with the popular and legal discourse touched on in the two sections above, where creation science debates and science and religion considerations seem to invite enthusiastic forays into "folk epistemology" to derive the essences of science and religion.

"Antiscience Studies"

A similar example of creation science debates being used to nurture polarized essentialist images of science can be found in attacks on science studies following in the wake of the so-called "science wars." Probably the most widely known example of this style of attack can be found in the work of Norman Levitt in his *Prometheus Bedevilled* (Levitt, 1999). I will quote liberally from this text to help convey to the reader the passion with which these claims are made:

> It is interesting, if dismaying, to note how far the attitudes and doctrines of the postmodern cultural left flatter the interests of creationists and similar groups on the cultural right
>
> (Levitt, 1999, p. 183)

> In reality, postmodern cultural anthropology is a bare inch away from being an outright ally of the creationist movement. Is it too alarmist to suggest as postmodern doctrine shifts in coming years, a syncretic alliance might develop between principled cultural relativists and fundamentalist anti-science?
>
> (Levitt, 1999, p. 186)

Levitt's critique is echoed by scientists such as Barry Palevitz:

> In the perpetual skirmish between science and religion, biological evolution is a contentious battleground. Despite a series of legal victories for science, which many thought or hoped would be the final nails in the coffin of creationism, hostilities still flare. Maybe it's millennialism. Perhaps it reflects unease with modern science and technology. Vigorous antiscience rhetoric coming from the humanities in the guise of postmodernism, some of it antievolution in flavor, may be fueling the fire. [*Footnotes removed from original quote*]
>
> (Palevitz, 2003, p. 171)

Palevitz also extends his "moral panic" into the broader intellectual terrain of debates about biological determinism, linking critics of biological determinism to creationists in their supposed shared fear of "honest objective inquiry."

> Perhaps the most pernicious message of creationists is this: If only we deny it, evolution will go away. If enough people are convinced, nature will follow suit. The radical postmodernists have a similar fantasy: If their objections to biological determinism (essentialism) are loud enough, perhaps genetic influences on human behavior will disappear. This kind of message should send shivers down the spine of anyone interested in honest objective inquiry. [*Footnotes removed from original quote*]
>
> (Palevitz, 2003, pp. 176–177)

For Levitt, the threat of creation science is linked to promoting a model of science education where science studies need to be excluded:

> It hardly needs stating that the climate currently prevailing in the science studies community is openly hostile to anything resembling the conventional notion of scientific literacy. To these partisans, what the general public should know about science is how wrongheaded and deluded scientists are and how insupportable it is for scientists to claim to know what, in fact, they really do know. For the moment, I can recommend no better antidote to the pretensions of science studies than well-informed scorn.
>
> (Levitt, 1999, p. 189)

What is interesting is that alongside their scorn for science studies is an expressed desire of the need for the curriculum to teach science students some kind of philosophy of science, albeit one purpose-built to only celebrate science's fundamental progressive rationality:

> We teach facts and principles in our science classes, but apparently not enough about the philosophical underpinnings of the scientific process and the way that scientists view the natural world.
>
> (Palevitz, 2003, p. 176)

Levitt and Palevitz's wish is to save science from creation science (and science studies) by monopolizing, for scientists like themselves, the right to comment and shape curricula on matters to do with science and society. They hope to do this by picking and choosing from the philosophical perspectives in science studies they like but never engage in scholarly dialogue with the perspectives they dislike. Such an approach hardly represents good social science scholarship. But in practical terms, even with a version of academic science studies devoid of social constructivist and postmodern tendencies, Levitt and Palevitz's simplistic essentialist folk epistemology would still be unlikely to "solve" the challenges they perceive. They would still be left adjudicating the differences between scientists such as Gould and Dawkins, and philosophers (with limited sympathies for social constructionism) such as Ruse and Laudan. Their polemical approach leaves them poorly equipped to do this.

Conclusion

Scholarship in the field of science studies (broadly defined) over the last twenty to thirty years has provided numerous philosophical perspectives and historical and contemporary case studies of science and technology. Echoing my earlier discussion of the work of Brooke, it is in this diversity and care for empirical detail that there is potential to avoid the tendency for creation science debates to encourage unnecessary polarization of perspectives and confused intractable debate over epistemological "essentialisms."

The diversity of perspectives offered by contemporary science studies does present some challenges for incorporation into the curriculum. There have been disagreements about what version of science studies should be adopted (Mathews, 1994), and some critics have suggested that the diversity of perspectives on offer means that the exercise should be abandoned (Shamos, 1995). The degree to which this diversity is an obstacle is nevertheless largely prefigured by perceptions that there is a need to have an essence of science taught in the curriculum, and ignorance, such as that displayed by Levitt and Palevitz, of how far many perspectives in science studies, across the relativist/realist continuum, have moved away from defining a single essence, ethos, or method of science. In a thoughtful discussion, John Rudolph reviews this question:

> It seems that much of the difficulty many see in incorporating the nature of science into the science curriculum resides in the particular manner in which the practice of science has been historically described. Such accounts have been concerned primarily with establishing some universal characterization of science into which the work of any given scientist in any discipline, past or present, might easily fit. This work has been motivated largely by the desire of scholars to provide a philosophical metascientific justification for the privileged place in society that science currently enjoys ... a goal

that, in many ways, overreaches the more circumscribed goals of science education.
[*Footnotes removed from original quote*]

(Rudolph, 2000, p. 404)

Rudolph notes two strategies that have been proposed in different contexts as ways to help overcome the lack of a consensus over a universal definition of science in science studies:

(1) *Try to abstract from competing views about the nature of science some more general lower-level statements about science from which educators may draw a consensus.* This strategy, argues Rudolph, can encounter problems if all it means is the adoption of trivial and very general statements that can be "noted" and then not be integrated in a meaningful way into specific teaching contexts.

(2) *Allow students to engage with competing philosophical perspectives.* Here, problems can be encountered on fairly pragmatic grounds: to engage in such an exercise normally presumes some degree of prior competence in science, which students simply may not have. This approach is also likely to be time consuming and begins to look like the promotion of science studies as a subject in its own right: a useful gesture, but something that represents a significant move from the initial aim of arriving at the best way to incorporate science studies into not replacing science education.

Rudolph suggests an alternative approach that emphasizes the investigation of science as "local practice." Rather than seek to first identify the ultimate universal goals of science, students could be encouraged to consider the cognitive goals, models, and methods of justification and argumentation of various scientific disciplines and clusters of disciplines. Students can begin to identify the different practices and cognitive styles manifest in various physical, chemical, and biological sciences. Insights from science studies can be used to help inform these discussions.

Going beyond Rudolph, the approach he supports could be augmented by inviting discussion of the social impacts of various sciences, considering the aims of particular disciplines, the norms of practitioners and practical outcomes again at a more immediate practically grounded level (Albury, 1983; Roth,McGinn and Bowen,1996). Considerations of philosophical metanarratives surrounding the broader legitimacy of science, and the uses to which this legitimacy has and should be put, involves more specialist history, sociology, and philosophy of science: concerns better dealt with in an expanded role for science studies in the social science/humanities curriculum. Arguably, one of the current obstacles to a more informed philosophical discussion of the politics and ethics of science is the prevalence of naïve realist philosophies of science that are appended to science classes with very little plausible link to the content of the sciences being taught. Ideally, if students have acquired a prior clearer understanding of features of local scientific practice, this will provide them with a better starting point to develop a better appreciation of the intricacies of epistemology science and ethics (Milne and Taylor, 1998).

The position I am advocating is compatible in many respects with some of the themes developed by Collins and Pinch in their popular science studies text *The Golem* (1993). In keeping with a body of literature that has investigated public perceptions of science in controversial areas (Irwin and Wynne, 1996), they observe that unsustainable metanarratives about the nature of science, apart from being historically and empirically unsustainable, can actually generate conflict by creating cognitive dissonance when experts, particularly in controversial settings, expose the slippage between the ideals of scientific practice and the practicalities and craft skills involved in scientific practice. These problems of public deconstruction of science have been a persistent theme hovering in the background of creation science debates. Because many creation science claims themselves are difficult to sustain empirically, it has been easier for creation science proponents to focus instead on exploiting the artificiality of ideal images of science so as to expose inconsistencies in mainstream scientific claims in biology. Holding evolutionary biology accountable to ideal canons of scientific method may leave biological claims whose cognitive and instrumental value is easy to justify according to canons of locally situated disciplinary practice open to unrealistic criticism. Critique of the social implications of such practices on the basis of such epistemological metanarratives is also likely to make little contact with the realities of science and public policy.

Rather than being drawn into answering fundamental issues of epistemology in the science classroom, creation science and intelligent design could be more constructively framed as "local knowledge" containing some elements loosely associated with contemporary biology. Within this more pragmatic framework, it would be relatively easy to assess the aims, cognitive style, modes of justification, and general (lack of) instrumental value of such claims. It would be unlikely that such claims would warrant much time in the science curriculum.

Broader religious/epistemological metanarratives associated with these claims would still be worth considering, but would be dealt with more constructively in the social science/humanities part of the curriculum. To be balanced, similar constraints could also be applied to assessing the more speculative biological determinist and scientistic elements of claims made by scientists such as Dawkins.

In all, if we are to capture the positive educational potentials of creation science debates, we need to reverse the trend for popular scientific discourse to become excessively preoccupied with producing simplistic folk epistemologies of science. These folk epistemologies often have at best tenuous links to academic science studies and tend to enhance confusion and polarized understandings of science and religion. Discussion of epistemologies of science are valuable but are likely to be more fruitful when they take place in scholarly contexts where there is some sense of longer-standing debates in science studies coupled with some prior-grounded appreciation of science as practice. Engagement in endless battles over folk epistemologies of science, such as those documented here, should not be allowed to direct energy in popular scientific discourse and science education away from engagement in the more important immediate project of encouraging a better appreciation of the actual practices of the sciences and their immediate social implications.

References

Albury, R. (1983). *The politics of objectivity*. Geelong: Deakin University Press.
Barnes, B. (1982). *Thomas Kuhn and social science*. New York: Columbia Press.
Biles, J. (2000 Oct 1). 'Inherit the Wind' more timely than ever' *LJWorld.Com*. Lawrence Kansas <http://www.ljworld.com/section/evolution/story/28410>
Brooke, J. H. (1991). *Science and religion: Some historical perspectives*. Cambridge: Cambridge University Press.
Brooke J. H. & Cantor, G. (1998). *Reconstructing nature: The engagement of science and Religion*. Oxford: Oxford University Press.
Cole, S. (2001). *Suspect identities: A history of fingerprinting and criminal identification*. Cambridge: Harvard University Press.
Collins, H. & Pinch, T. (1993) *The golem: what everyone should know about science* Cambridge: Cambridge University Press.
Dawkins, R. (2003). You can't have it both ways: irreconcilable differences. In P. Kurtz, (Ed.) *Science and religion: are they compatible?* (pp. 207–212). New York: Prometheus Books.
Dayton, L. (1997). Fraud or copyright, science or faith—it's simply about basic beliefs. *The Sydney Morning Herald*, April 8, 6.
Edmond, G. & Mercer, D. (1996). What judges should know about falsification. *Expert Evidence*, 5, 29–42.
Edmond, G. & Mercer, D. (1998).Trashing Junk Science. *Stanford Technology Law Review*. 3. <http://stlr.stanford.edu/STLR/Articles/98_STLR_contents_f.htm>
Edmond, G. & Mercer, D. (1999). Saving science: creating (public) science in the Noah's Ark case. *Public Understanding of Science*, 8, 317.
Edmond, G. & Mercer, D. (2002). Conjectures and exhumations: citations of history, philosophy and sociology of science in US Federal Courts, *Law and Literature*, 14, 309–366.
Geiryn, T. (1998). *Cultural Boundaries of Science: Credibility on the Line*. Chicago: Chicago University Press.
Geiryn, T., Bevans, G., & Zehr, S. (1985). Professionalization of American scientists: public science in the creation/evolution trials, *American Sociological Review*. 50, 392.
Gould, S. J. (1977). *Ever Since Darwin: Reflections in Natural History*. New York: Norton.
Gould, S. J. (1999). *Rock of Ages*. New York: Random House.
Gregory, J. & Miller, S. (1998). *Science in public: Communication, culture and credibility*. New York: Plenum.
Irwin, A. & Wynne, B. (1996). (Eds) *Misunderstanding science? The public reconstruction of science and technology*. Cambridge: Cambridge University Press.
La Follette, M. (1983). Creationism in the news: Mass media coverage of the 'Arkansas Case' in M. La Follette (Ed.) *Creationism, science and the law: The Arkansas case*. (pp. 189–208), Cambridge MA: MIT Press.
Larson, E. (1997). *Summer for the Gods: The Scopes trial and America's continuing debate over science and religion*. New York: Basic Books.
Laudan, L. (1982). Science at the Bar: Causes for Concern. *Science Technology & Human Values*, 7, 61.
Levitt, N. (1999). *Prometheus Bedevilled: Science and the contradictions of Contemporary Culture*. London: Rutgers University Press.

Locke, S. (1994). The use of scientific discourse by creation scientists: some preliminary findings. *Public Understanding of Science*, 3, 403.

Milne, C. & Taylor, P. (1998). Between a myth and a hard place: situating school science in a climate of cultural reform. In W. Cobern (Ed.), *Socio-Cultural perspectives on Science Education: An International Dialogue* (pp. 25–48). Dordrecht: Kluwer.

Mathews, M. (1994). *Science teaching: The role of history and philosophy of science*. New York.: Routledge.

Mulkay, M. (1979). *Science and the sociology of knowledge*. London: Allen and Unwin.

Nelkin, D. (1977). *Science textbook controversies and the politics of equal time*. Cambridge, MA.: MIT Press.

Nelkin, D. (2004) God Talk: Confusion between science and religion: Posthumous essay. *Science, Technology and Human Values*, 29, 139.

Palevitz, B. (2003). Science versus religion: a conversation with my students. In P. Kurtz (Ed.) *Science and religion: are they compatible?* (pp. 171–179). New York: Prometheus Books.

Pockley, P. (1997a). Creationism 'Ark' trial opens in Australia, *Nature*, 386, 529.

Pockley, P. (1997b). Science, fundamentalism or Noah's ark? *Search*, 28, 147–149.

Quinn, P. (1984). The philosopher of science as expert witness. In J. Cushing, C. F. Delaney & G. Gutting (Eds) *Science and Reality: Recent Work in the Philosophy of Science* (pp. 32–53). Notre Dame, IN.: University of Notre Dame Press.

Roth, W. M., McGinn, M. K. & Bowen, G. M. (1996). Applications of science and technology studies: effecting change in science education, *Science, Technology & Human Values*, 21, 454–484.

Rudolph, P. (2000) Reconsidering the 'nature of science' as a curriculum component, *Journal of Curriculum Studies*, 32, 403–419.

Ruse, M. (1982). *Darwinism Defended: A Guide to the Evolution Controversies*. California: Benjamin Cummings.

Ruse, M. (1996a). Witness testimony sheet, McLean v. Arkansas. In M. Ruse (Ed.) *But is it Science? The Philosophical Question in the Creation/Evolution Controversy* (pp. 302–306). New York: Prometheus Books.

Ruse, M. (1996b) Prologue: a philosopher's day in court. In M. Ruse (Ed.) *But is it Science? The Philosophical Question in the Creation/Evolution Controversy* (pp. 13–35). New York: Prometheus Books.

Scott, E. (1997). Creationism, ideology and science. In P. Gross, N. Levitt & M. Lewis (Eds), *The Flight from Science and Reason* (pp. 505-523). London: Johns Hopkins Press.

Scott, E. (2003). The science and religion movement: an opportunity for improved public understanding of science. In P. Kurtz (Ed.) *Science and Religion: Are they Compatible?* (pp. 111–116) New York: Prometheus Books.

Shamos, M. (1995). *The myth of scientific literacy*. New York: Rutgers University Press.

Shermer, M. (1997). *Why people believe weird things: Pseudoscience, superstition, and other confusions of our time*. New York: Freeman.

Wertheim, M. (1999). A rock and a hard place, *The Australian*, October 13, 22.

Wilson, D. (2002). The historiography of science and religion. In G. Ferngren (Ed.) *Science and Religion: A Historical Introduction*. (pp. 13–29). Baltimore: The Johns Hopkins University Press.

Wilson, T. (2003). Evolution, creation, and naturally selecting intelligent design out of the public schools, *Toledo Law Review*, 34, 203.

Young, R. M. (1985). *Darwin's metaphor: Nature's place in Victorian culture*. Cambridge: Cambridge University Press.

How Not to Teach the Controversy about Creationism

Robert T. Pennock

> *I think that what [the Kansas Board of Education] would have liked to get, if they'd had their druthers, is the kind of thing that I advocate: you present the dissenting case [against evolution] on equal terms, and you teach the controversy.* (Johnson, 1999)

Introduction: Teach the Controversy?

Kansas just can't get a break. In 1999, the state became an international laughingstock when creationists on the State Board of Education, led by Steve Abrams, gutted what would have been a model science curriculum, removing the theme of evolution as well as mentions of the Big Bang and the geological timescale (Pennock, 1999b, 2000). These board members and the creationist groups that assisted them seemed to confirm every stereotype of Kansas as an ignorant backwater. The creationists lost their majority on the Board in the next election when sensible Kansans made their voices heard, and the standards were righted. But freedom—in this case, freedom from the tyranny of ignorance—demands eternal vigilance, and science defenders were too quick to rest on their laurels. Another election passed and in 2005 Steve Abrams and his allies were at it again.

In the 1999 round, Abrams initially conspired with the Creation Science Association of Mid-America.[1] Intelligent design creationism (IDC) lobbyists from the Discovery Institute (DI) quickly jumped in to assist. In "Teaching the Origins Controversy: A Guide for the Perplexed," DI Fellow David DeWolf discussed how intelligent design can be taught as an "alternative," "competing" theory, and he offered the resources of the Discovery Institute—ID experts who were purportedly "real scientists" doing "cutting-edge research," legal support, and planned curriculum materials and in-service workshops. Speaking for DI, DeWolf wrote "the reason we think that a biology teacher would be wise to present intelligent design along with Darwinism is that it is a necessary corrective to the impression frequently given that Darwinism is the only scientific theory of biological origins" (DeWolf, 1999).

---

[1] Board members denied that creationists had any hand in the revisions, but Kansas Citizens for Science documented that Tom Willis, president of the Creation Science Association of Mid-America, had secretly authored the changes (Krebs and Case, 1999).

In the latest round in Kansas, many of the same players are involved, but this time Abrams and the DI have had to change their approach. The ID movement has seen a series of failures after trying to get their view mandated, and has since been in retreat. In Ohio, for example, IDCs working with supporters on the State Board of Education tried for two years to get intelligent design mandated in the model curriculum, but after strong opposition made it clear this would fail, they had to withdraw and tried a "compromise" proposal in March 2002 that would allow teachers to tell students about "the scientific controversy over Intelligent Design" (Princehouse, 2006). Although one can find this notion in earlier IDC writings, since the defeat in Ohio they have mostly dropped attempts to insert ID explicitly and instead promote this backdoor strategy, adopting it as their slogan, as announced in an interview at the time: "Our slogan to the press is, Teach the controversy, said [Phillip] Johnson, widely regarded as the father of the modern intelligent design movement" (Stephens, 2002). As can be seen in the epigraph from Johnson that heads this article, their original goal was to have their dissenting view taught "on equal terms" with evolution. However, in their recent efforts, they recommend that their supporters not directly mention ID but rather introduce policies to teach the "strengths and weaknesses" of evolution or the "arguments for and against" evolution and to open the door to the inclusion of unspecified "alternative theories."

Discovery Institute ID advocate Jonathan Wells' book *Icons of Evolution* is often put forward as a model. In Texas in 2003, for example, IDCs lobbied the Board of Education to reject biology textbooks that did not include what they said were the "weaknesses" of evolution, such as the kinds of challenges made by Wells. Compare these with the 2004 case in Pennsylvania, where a creationist-dominated school board in Dover adopted a policy requiring that "Students will be made aware of gaps/problems in Darwin's theory and of other theories of evolution including, but not limited to, Intelligent Design" (Directors, 2004). This case is especially revealing about the revised strategy, in that the Discovery Institute did not support the policy. Richard Thompson, head of the pro-creationism Thomas More Law Center,[2] which defended the school district in the *Kitzmiller v. Dover* case involving the ID policy, was publicly angered when a DI representative claimed that they had never advocated including ID. Thompson held up documents from DI and pointed out places in which they had done just that, chiding them for "intellectual dishonesty" and for abandoning those who had taken their advice (Sulon, 2005).

The ID terminology had come to the fore in about 1987, when the US Supreme Court ruled that teaching "Creation Science" was unconstitutional. Creationists had to quickly distance themselves from that language. Although it

---

[2] The Thomas More Law Center's slogan is "The Sword and Shield of People of Faith" and it takes on cases such as the Dover on for free as part of its mission, as stated on its web site, to "protect Christians and their religious beliefs in the public square."

followed the same general strategy as Creation Science of claiming to be scientific, ID was put forward more narrowly—in what they called the "wedge" strategy—so it would have a better chance of pushing through the wall of separation between church and State that its predecessor had failed to breach. Because ID has now similarly become clearly recognizable as stealth religion, it no longer serves their purpose to lobby directly for it under that terminology. Thus, we now observe this new strategy, which is to sharpen the wedge still further. For the purposes of this paper, I will focus mostly on two main statements of this new ID approach. The first comes from a series of newspaper op-editorials starting with one titled "Teach the Controversy" (Meyer, 2002) by DI CSC director Stephen Meyer and including ones co-written with DI Fellow John Angus Campbell (Meyer and Campbell, 2004a, 2004b; Campbell and Meyer, 2005; Meyer and Campbell, 2005).[3] The second comes from a recent video from ColdWater media and the DI called "How to Teach the Controversy over Darwin Legally." As we will see, this latest trick of the ID movement is just another attempt to disguise their sectarian religious view.

Mirrors and Smoke

A common magician's stage trick with mirrors is to use multiple reflected images to make something appear to be what or where it is not. ColdWater Media, which produced and distributes the How-To video, has produced three other ID films—about half of its current catalog—but this video is really a Discovery Institute show from beginning to end. Of the half-dozen people interviewed in the video—Stephen Meyer, John West, David DeWolf, Doug Cowan, Scott Minnich, and Bruce Chapman—only Chapman is identified as being affiliated with DI. But this is misleading as all except Cowan is either an administrator or Fellow of DI.[4]

The video begins by saying local people are confused in thinking that teaching antievolution is unconstitutional. It warns that people should expect intimidation and threats of lawsuits from the ACLU, but goes on to say that it is wrong to worry about possible expense for the school district and a loss in

---

[3] Meyer's solo op-ed was occasioned by the Board of Education controversy in Ohio in March 2002. The four joint pieces appeared in February 2004, December 2004, January 2005, and March 2005. The first of these addresses a Georgia case where a superintendent tried to remove the word "evolution" from the science standards. The second and third focus on the Dover, Pennsylvania case. The last does not mention any particular controversy. Although they all have different titles and Meyer lists them all separately on his CV, they are all nearly identical. The two most similar ones appear superficially different because of very different paragraph breaks, but differ only in a dozen or so sentences and in the order of the authors. The two most different ones are still nearly two-thirds word-for-word the same. None mention that the material is previously published. I have previously found this unusual cut-and-paste self-plagiarism in other writings of Meyer and company (Pennock, 2004).

[4] Cowan is identified only as "a public high school teacher in a large urban school district." Even this is misleading; Cowan actually teaches in University Place, Washington, which describes itself as a suburban residential community of about 32,000 people.

court. Why? Because, claims the video, this is not religion but science. But is that really true?

In addition to point-man Johnson, there are a half-dozen other core leaders in the group whose names quickly become familiar as one reads the ID writings, including Michael Behe, William Dembski, Paul Nelson, Jonathan Wells, and Stephen Meyer. This group is representative in many ways of the "big tent" approach that the ID movement takes. The movement is a coalition of young earth creationists like Nelson and old earth creationists like Meyer. It includes evangelical Presbyterians like Johnson, evangelical Catholics like Behe, and evangelical Moonies like Wells. They and other creationists in the movement are united in desire to replace Darwinian evolution with special creation, temporarily setting aside internal differences to fight together under a banner of "Mere Creation" against their common foe (Pennock, 1999a).

As they portray their movement in public forums, intelligent design is not about the Bible. It is not religion but science, they claim. In an op-editorial in the New York Times, Michael Behe wrote that "the theory of intelligent design is not a religiously based idea ... intelligent design itself says nothing about the religious concept of a creator" (Behe, 2005). Such brazen statements are astonishing to anyone who is familiar with the ID literature. In other settings, such as when they are speaking to their rank and file supporters, they are more forthright. In an interview to an evangelical Christian audience on the American Family Radio show, Johnson explained their strategy and primary goal:

> Our strategy has been to change the subject a bit so that we can get the issue of intelligent design, which really means the reality of God, before the academic world and into the schools.
>
> (Johnson, 2003).

Further details of the ID perspective and strategy were revealed in a leaked internal fundraising statement from the Discovery Institute known as the Wedge document. The document discusses their plans to achieve their governing goals, namely "To defeat scientific materialism and its destructive moral, cultural, and political legacies" and "To replace materialistic explanations with the theistic understanding that nature and human beings are created by God" (Discovery Institute, 1999). It states that their aim is to make ID an accepted alternative and eventually the dominant view in science as the key to achieving this goal. Evolution and materialism are taken to be the basis of cultural evils, and the cure, IDCs believe, is acceptance of the Biblical view of man as created in the image of God. As Barbara Forrest and Paul Gross put it in their detailed study of the Wedge movement, intelligent design is creationism's Trojan horse (Forrest and Gross, 2003).

In its How-To video, DI recommends that antievolutionists "keep focus off religion" when they lobby for teaching criticisms of evolution. The recommendation itself is revealing of the religious nature of the project; one would never think of mentioning religion, let alone of warning against talking about it, if this were a scientific debate. But of course this is a religious, not a scientific issue. Really, DI is recommending that its supporters be deceptive and try to keep that under wraps.

The video is similarly misleading in the way it describes the Supreme Court ruling, which it claims was just against "biblically based creationism." DeWolf says that "the Supreme Court held that that method of approaching the subject, that is, starting with the Bible, and essentially looking for evidence that would confirm the statements in the Bible that that wasn't a scientific method that was comparable to the teaching of evolution" (Discovery Institute, 2005). In fact, the Louisiana Creationism act that was in question did not mention the Bible, but claimed, in the same way that ID does now, that the schools should teach "the scientific evidences for [creation or evolution] and inferences from those scientific evidences" (Edwards v. Aguillard, 1987). Seeing through the disguise, the court rejected the claim that Creation Science was science.

Behind the Curtain of "Teaching the Controversy"

The strategic silences in the Discovery Institute film are deafening. The term "intelligent design" is never mentioned even once, though its content permeates the video. This reminded me of a talk I heard Dembski give to high school teachers on how to include ID, where at the end he pointed out to the audience that he had given the whole story of design without using "the G-word."

The film puts forward high school teacher Doug Cowan as its model for teaching the controversy, so it will be informative to see what Cowan actually includes in his class. Cowan described some of what he does in an article that appeared about a month before the film was released. He runs down the same checklist of points that the film does, claiming that the Supreme Court allows what he does, that what he calls a "Senate addendum" to the No Child Left Behind Act endorses the approach, that teaching the controversy is in line with state requirements to teach critical thinking, and that it is a "freedom of speech issue." (We shall consider these in more detail below.) But he also writes:

> My superintendent asked me to stick to the adopted curriculum—which does not include intelligent design theory—and I've done so. However, I have retained the freedom to mention intelligent design theory to curious students as another viewpoint used to explain life and its diversity.
>
> (Cowan, 2005).

He tells students that "contrary to their large and monolithic biology textbook, some highly credentialed scientists insist that there are limitations to Darwin's theory." He tells them that "some current biology textbooks contain widely discredited evidence for Neo-Darwinism." And he tells them that "even respected scientists have peddled fraudulent evidence in defense of a pet scientific dogma." Even with this small sample, it strains credulity to believe that Cowan is presenting the facts even in a "neutral" manner, let alone in a responsible manner. When one looks at the few specifics he mentions of what he teaches, it becomes clear that it is Cowan who is peddling dogma, bringing up the same false and misleading charges that ID and Creation Science have always done.

Cowan says he teaches about the Piltdown Man, a 1912 hoax of an apparent "missing link" that purportedly was uncovered in a gravel pit in southern England. This old story is a favorite of creationists. But one wonders whether Cowan then goes on to explain that it was scientists like Oxford University professor E. T. Hall who eventually debunked the Piltdown hoax using X-ray fluorescence techniques, which revealed that the bones had been stained with potassium dichromate to make them appear fossilized, after later archaeological discoveries raised suspicions about it. The Piltdown case is actually a wonderful example of the self-correcting nature of science—even clever hoaxes can be found out. And does Cowan present the dozens of legitimate hominid and other transitional fossils uncovered in the subsequent century? Indeed, why bring up a century-old hoax at all if not to cast doubt on scientific evidence that current scientists present?

Cowan says he teaches his students about "Ernst Haeckel's faked embryo drawings," which made early stages of embryonic development in different species appear more similar than they really are to support his idea that an organism's embryological development recapitulated its evolutionary phylogeny. This case is one of the main criticisms of evolution featured in ID creationist Jonathan Wells's book *Icons of Evolution*, which has been a blueprint for supposed problems with evolution (Wells brings up Piltdown Man too). The Haeckel case dates even further back, originally to one particular illustration from 1874. It is notable that other scientists at the time identified Haeckel's inaccuracies and charged him with falsifying science. Haeckel defended his schematic drawings, noting that all such "diagrammatic" figures are necessarily inaccurate, but he did modify them in later editions to make them a bit more precise. Still, it is true that his inaccurate drawings continued to be reproduced even in a few current textbooks, so this one case where there is a problem that had to be remedied. But, again, one wonders whether Cowan highlights how it was scientists, most recently and forcefully Michael Richardson and colleagues, who called attention to Haeckel's inaccuracies. (Wells draws his information primarily from Richardson.) And does Cowan fully describe how the embryological evidence as understood today fits with evolution? Learning that their work was being misused by IDCs, Richardson and colleagues wrote:

> Our work has been used ... to attack evolutionary theory, and to suggest that evolution cannot explain embryology. We strongly disagree with this viewpoint. Data from embryology are fully consistent with Darwinian evolution.... On a fundamental level, Haeckel was correct: All vertebrates develop a similar body plan (consisting of notochord, body segments, pharyngeal pouches, and so forth). This shared developmental program reflects shared evolutionary history. It also fits with overwhelming recent evidence that development in different animals is controlled by common genetic mechanisms
>
> (Richardson, Hanken *et al.*, 1998).

Cowan goes on to say that he also presents the reputable evidence for evolution, listing "genetically altered fruit flies, the evolution of antibiotic resistance in bacteria and insecticide resistance in bugs, how breeding programs change domestic species, and how oscillating climates affect the beak size of certain kinds of finches." However, these carefully selected items and the way he describes them shows that even here he is following Wells' blueprint. He says that these demonstrate only that organisms are capable of change over time, but then asks students whether such microevolutionary changes can be extrapolated to explain macroevolutionary change, which he inaccurately defines as "evolution from one type of creature to a fundamentally different kind" (Cowan, 2005). In standard biological definitions, there is an intentionally fuzzy line between microevolution and macroevolution; the former refers to changes within a species and the later to somewhat greater changes that result in the formation of new species. It used to be that creationists would not accept even microevolution, but now virtually all creationists accept it, but draw the line there or above closely related species. Cowan appears to be following the standard creationist line. Indeed, Cowan further says that he calls the reputable evidence into question by "dissect[ing] these evidences using recent discoveries that have raised important questions among evolutionary biologists" without saying what these might be (Cowan, 2005). He uses the creationist term "evidences," rather than the standard scientific term.

Thus, although ID was, strategically, not mentioned by name in the film, Cowan includes it both explicitly and implicitly throughout his class presentation. In practice, "teaching the controversy" is equivalent to teaching ID. Unsurprisingly, this is just another Trojan horse.

The ID Mirror Cracks

Meyer claimed that schools should teach the controversy because "First, honest science education requires it" (Meyer, 2002). This statement is preposterous on its face. More than 70 professional scientific organizations have issued statements opposing intelligent design and testifying to the evidential strength and central explanatory importance of evolution. Might all these scientists be misinformed? Is not evolution a "theory in crisis" that is about to collapse, as IDCs claim? Might there not be new evidence for intelligent design that challenges evolution? With hundreds of articles published on evolution each year in peer-reviewed science journals, there is a huge body of evidence that backs up the scientific consensus, so this seems unlikely, to put it mildly. But do IDCs have any evidence on their side?

In his op-ed, Meyer also cited an annotated bibliography of forty-four "peer-reviewed scientific articles" that purportedly raised "significant challenges to key tenets of Darwinian evolution" that he and Wells presented to the Ohio Board of Education when they were lobbying it to teach the controversy. Was this at last the presentation of peer-reviewed intelligent design research? Based on how they presented it, many ID supporters and even some journalists assumed so, and described it as such. But a closer look revealed quite a different picture. The National Center for Science Education (NCSE) sent a questionnaire to the authors of every article in the DI Bibliography, asking whether they took their research to provide scientific evidence of intelligent design. The results were not surprising:

> None of the 26 respondents (representing 34 of the 44 publications in the Bibliography) did; many were indignant at the suggestion. For example, Douglas H. Erwin answered, "Of course not—[intelligent design] is a *non sequitur*, nothing but a fundamentally flawed attempt to promote creationism under a different guise. *Nothing* in this paper or any of my other work provides the slightest scintilla of support for 'intelligent design'. To argue that it does requires a deliberate and pernicious misreading of the papers."... Several respondents even went so far as to say that their work constituted scientific evidence *against* intelligent design
>
> (NCSE Staff, 2002).

NCSE also asked whether the authors believed that their articles challenged evolution. Again, the answers were as expected:

> None of the respondents to NCSE's questionnaire considered their work to provide scientific evidence against evolution. David M. Williams, for example, simply remarked, "No, certainly not. How could it possibly?"
>
> (NCSE Staff, 2002).

NCSE posted an article giving the results and going on to discuss what they discovered in examining the content of the articles in the DI Bibliography and the way that DI's annotations summarized them. More than half of the DI summaries were rejected by the original authors as dishonest, misleading, inaccurate, and tendentious or the like. I will not take the time to review the content analysis further here, but the upshot is that this was yet another case of the standard creationist practice of quote-mining and misrepresenting research for their own purposes (NCSE Staff, 2002). For instance, many of the articles were devoted to extending evolutionary theory, but such research is hardly a "challenge" to the theory in the sense that IDCs imply. Evolutionary theory has been greatly developed and extended in the century and a half since Darwin, so it hardly even makes sense to speak just of "Darwinism" or "Darwin's theory" in the way that creationists do.

Meyer also mentions a statement by 100 scientists "questioning the creative power of natural selection" that the Discovery Institute organized in response to a 2001 public television series on evolution. The title of the statement was "Scientists Dissent from Darwinism," which most people would interpret as a broad rejection of evolution. However the statement read only: "We are skeptical of claims for the ability of random mutation and natural selection to account for the complexity of life. Careful examination of the evidence for Darwinian theory should be encouraged." Rather than a broad dissent, this wording was very narrow, omitting any mention of the evolutionary thesis of common descent, human evolution, or any of the elements of evolutionary theory except for the Darwinian mechanism, and even that was mentioned in a very limited and rather vague manner. I have done research demonstrating the power of the Darwinian mechanism to produce complexity of just the sort that IDCs claim is impossible (Lenski *et al.*, 2003) and am a staunch defender of the broad applicability of the process, yet not even I would claim that mutation and natural selection alone account for every aspect of life's complexity. So this is not really a radical statement.

The National Center for Science Education wrote to all the signatories asking them whether they thought that living things shared common ancestors and whether humans and apes shared common ancestors. In an interview on *The Science Show*, Eugenie Scott, director of NCSE, described the response:

> [W]e got back, within the first couple of hours, two or three of these people, the prominent scientists saying, "Oh yeah, evolution took place and yeah, I have no problem with humans evolving from apes, but I just don't think natural selection is this really all powerful mechanism, it can't explain the origin of life." And then of course we got 'no responses' because what we surmise happened is, the Discovery Institute found out about this email and clamped down and said "Don't answer anybody back". So we have evidence that at least *some* of the more knowledgeable scientists did not interpret this DI100 statement the way that it was intended to be interpreted by the general public
>
> (Williams, 2003).

There were also some other curious things about the list she mentioned. In fine print at the bottom, it was noted that the signatories were "identified by institution of degree or present position." That kind of "either/or" label is odd and makes one wonder whether there might be a bit of misdirection going on here too. Sure enough, a little research showed that people who worked for Probe Ministries or the Institute for Creation Research, for example, were listed instead by where they got their degrees.

In comparison, one might consider another statement, one put together by the NCSE, which read as follows:

> Evolution is a vital, well-supported, unifying principle of the biological sciences, and the scientific evidence is overwhelmingly in favor of the idea that all living things share a common ancestry. Although there are legitimate debates about the patterns and processes of evolution, there is no serious scientific doubt that evolution occurred or that natural selection is a major mechanism in its occurrence. It is scientifically inappropriate and pedagogically irresponsible for creationist pseudoscience, including but not limited to "intelligent design," to be introduced into the science curricula of our nation's public schools.

The signatories to this broader statement totaled over 200 within a couple of weeks and number 693 as of January 2006. There is one other feature about this statement that distinguishes it from the Discovery Institute list and other such lists that have been promulgated over the years by creationists—the scientists who signed the NCSE statement are all named Steve. (Actually,

Stephen or Stephanie or other cognates were also accepted. The name Steve was picked in honor of the late Stephen Jay Gould.) With about 1% of the US population having this given name, the number of signatories corresponds to over half a million scientists.

Returning now to Meyer's claim about whether teaching the controversy in their sense is "honest" science education, the evidence has shown that it is not. The pattern of misrepresentation and dishonesty that was demonstrated in the cases above from the Discovery Institute is all too typical of creationist attacks on evolution.

Clearing the Smoke

The other arguments that IDCs bring up amount to mere smoke-blowing and may be dispensed with quickly.

Is it true that constitutional law permits it?

We have previously shown how IDCs misrepresent the Supreme Court ruling against teaching creationism. Incredibly, IDCs sometimes claim that the court explicitly endorses their view:

> Interestingly, the court also determined that teachers have the right to teach students about "a variety of scientific theories about origins... with the clear secular intent of enhancing science education" [sic]
>
> (Campbell and Meyer, 2005).

Actually, Meyer and Campbell misquote this comment from the 1987 *Edwards v. Aguillard* decision. Noting that the courts have allowed that it may sometimes be permissible to use the Ten Commandments in a class, the ruling mentions hypothetically that:

> In a similar way, teaching a variety of scientific theories about the origins of humankind to schoolchildren might be validly done with the clear secular intent of enhancing the effectiveness of science instruction
>
> (*Edwards v Aguillard*, 1987).

The statement is not a determination of the case per se, which overturned a state act that required balanced treatment of evolution and Creation Science, but rather is a caveat made in passing. Is this a real loophole in the law? It should not be seen as such. We have already seen how ID is advanced for a clear religious purpose and how teaching it is the very opposite of effective science instruction. Meyer and Campbell's op-editorials are misleading in focusing on a single line taken out of context, and ignoring the major finding of the court, which is that teaching creationism is unconstitutional since it lacks a secular purpose and violates the establishment clause. The key point, even in the passage quoted, is that legitimate criticisms and alternatives must be *scientific*. The central ruling of the Supreme Court was that the Act unconstitutionally endorsed religion "by advancing the religious belief that a supernatural being created humankind" (Edwards v Aguillard, 1987). That is exactly what intelligent design does. The bottom line is that ID is religious in just the same way that Creation Science was, so teaching it is unconstitutional for the same reason.

Is there really a federal mandate?

IDCs claim the imprimatur of what they call a "Senate addendum" to the No Child Left Behind (NCLB) act as a justification for teaching intelligent design (Cowan, 2005). They regularly quote the following: "[W]here topics are taught that may generate controversy (such as biological evolution), the curriculum should help students to understand the full range of views that exist [and] why such topics may generate controversy." They describe this as "authoritative report language" (Meyer and Campbell, 2005) of the Act or even as "federal education policy" (Meyer, 2002). This is highly misleading. What is the real story here?

The quoted line is part of compromise wording that had begun as a proposed amendment to the NCLB bill introduced by Senator Rick Santorum, a long-time ID supporter. Indeed, the original language of Santorum's amendment had been written by Philip Johnson himself (Applegate, 2001). However, the amendment was dropped from the final bill, and even the two-sentence watered-down language was relegated to a conference committee report. The quoted line does not have the force of law. Neither it nor any of the Santorum language nor even any mention of evolution appears in the NCLB act.

In a letter replying to a request for clarification regarding the status of the committee report, Congressman George Miller wrote:

> The law restricts the federal endorsement of curriculum, and the report language should not be construed to promote specific topics within subject areas. Congress recognizes that the teaching of the "full range of scientific views" should be encouraged, and such decisions are best left to the scientific community, rather than legislators
>
> (Miller, 2002).

Miller's letter goes on to emphasize the part of the compromise language that Meyer did not cite:

> The report language correctly describes science as a subject of "data and testable theories," different from "religious or philosophical claims." It is critical that the effort to narrow the achievement gap not be burdened with ideology regardless of subject matter.
>
> (Miller, 2002).

ID is not testable. It is not a scientific theory, but is rather religious ideology masquerading as one. Thus, it could not have properly been included on the basis of the compromise language even if it had been enacted.

What about academic freedom?

The DI How-To video speaks of defending the academic freedom of teachers who want to teach the controversy, and Cowan calls it a freedom of speech issue. Creation scientists have been making this specious argument for decades, and it has not improved in the ID version. Three court cases—*Webster v. New Lenox School District* in 1990, *Peloza v. Capistrano Unified School District* in 1994, and *LeVake v. Independent School District #656* in 2001—have already rejected different variations of the claim that teachers have a right to teach such material.

Independent of the legal issues, the appeal to academic freedom to include intelligent design is improper, given the nature of the issue. Creationists conveniently forget that with freedom comes responsibility, and academic freedom is no different. Academic freedom is not an open license to teach anything one wishes, but rather to teach—responsibly—material that is appropriate to one's discipline. Given the centrality of evolution in biology, teaching the controversy in the way that IDCs want would be the height of professional *irresponsibility* (Pennock, 2002).

Recognizing that evolution is "all but universally accepted in the community of scholars and has contributed immeasurably to our understanding of the natural world," the American Association of University Professors recently adopted a resolution stating that:

> [AAUP] deplores efforts in local communities and by some state legislators to require teachers in public schools to treat evolution as merely a hypothesis or speculation, untested and unsubstantiated by the methods of science, and to require them to make students aware of an "intelligent-design hypothesis" to account for the origins of life. These initiatives not only violate the academic freedom of public school teachers, but can deny students an understanding of the overwhelming scientific consensus regarding evolution
>
> (AAUP, 2005)

What about the polls?

Meyer and other IDCs often mention polls that indicate that a majority of Americans support for teaching evidence both for and against Darwin's theory of evolution. One common figure they cite is that 71% of Americans say they agree with the statement that "Biology teachers should teach Darwin's theory of evolution, but also the scientific evidence against it." Another is that 78% say they agree with the statement that "When Darwin's theory of evolution is taught in school, students should also be able to learn about scientific evidence that points to an intelligent design of life" (Zogby, 2001).

There are only a couple of things that need to be said about this. The first is to point out that the wording of these poll questions is misleading in that it improperly suggests that there is scientific evidence against evolution and for intelligent design. Neither is the case. However, an ordinary person answering a poll question would not be in a position to know this. The misleading wording is no surprise, given that the poll was commissioned by the Discovery Institute. The second is that poll results are irrelevant to the substantive question of what is proper to teach in a science class. One should teach science only in science classes, and scientific conclusions are not decided by a vote or a popularity contest.

Is this good pedagogy?

Finally, IDCs often say that they want to teach more about evolution and that teaching the controversy is good pedagogy. The DI How-To video makes this claim, but then just briefly mentions that the majority in science may be wrong and that students should also hear the minority view. Cowan claims that he "remains neutral" in the way he presents the material. He said that students appreciate this, noting one girl who liked it because "hearing the evidence for and against the theory gave her the freedom to weigh the evidences for herself" (Cowan, 2005). Students may like this now, but Cowan is doing them a deep disservice to suggest any sort of scientific neutrality regarding the evidence. Cowan closes his piece, as do Meyer and Campbell, with a line from Darwin that IDCs quote in half of their articles: "A fair result can be obtained only by fully stating and balancing the facts and arguments on both sides of each question." This, they always say, is what science is all about. Creationists quote this line to suggest that even Darwin would have supported including their view. But we have already seen why this is disingenuous. Evolutionary theory has been developing for over a hundred and fifty years since Darwin published the *Origin*, and it is absurd to suggest that one should continue to lay out the options neutrally as though it were still an open question. This would be equivalent to presenting the evidence for heliocentrism neutrally.[5] Indeed, it would be pedagogically irresponsible to misrepresent the findings of science by teaching in this way.

Conclusion: Teach Real Science

IDCs hope that "teaching the controversy" in the way that they propose will pass through the loophole they believe they have identified in the Supreme Court's rulings, because they claim it enhances the effectiveness of science education as a secular manner. They claim a right of academic freedom and say their view will pass the constitutionality test, that it is supported by a federal mandate, that opinion polls support the notion, and that doing so is good pedagogy. But, as we have seen, none of this is true.

Their political arguments are specious and their scientific and pedagogical claims are baseless. It is the very opposite of effective teaching to misrepresent the conclusions and methods of a discipline. Their declarations to the contrary notwithstanding, evolution is firmly established, and ID is not a real alternative

---

[5] Ironically, Tom Willis, the creationist leader who rewrote the Kansas standards in 1999, opposed not only evolution but also heliocentrism.

scientific view. Adopting their approach would misrepresent science and harm science education.

What should be educators' and citizens' response when IDCs lobby in Kansas and elsewhere to "teach the controversy"? We should respond with a slogan of our own: Teach only real science in science classes, not creationist pseudoscience!

The courts agree with this assessment. In December 2005, the *Kitzmiller v. Dover* case regarding the Dover, Pennsylvania ID policy was decided. During six weeks of detailed testimony, ID advocates, including Michael Behe and Scott Minnich, who are cited as their "scientific" authorities, presented their best arguments. Several of us who had the privilege of serving as the expert witnesses for the proscience side gave ours. The Discovery Institute submitted an amicus brief, attempting to rebut our evidence and arguments. But the judge was not convinced, finding, for reasons including but far from limited to those considered above, that teaching ID is unconstitutional in just the same way that Creation Science is. The Court said that "An objective observer would know that ID and teaching about 'gaps' and 'problems' in evolutionary theory are creationist, religious strategies that evolved from earlier forms of creationism" (*Kitzmiller v. Dover*, 2005, p. 18). The judge found that ID distorted and misrepresented the status of evolution by falsely suggesting that it was unreliable or on shaky ground (*Kitzmiller v. Dover*, 2005, p. 41, 84–86). And in a stinging repudiation of the central claims of the ID movement, the Court found that (1) ID is not science, (2) that it is a form of creationism, and (3) that it is inextricably religious in nature:

> [W]e have addressed the seminal question of whether ID is science. We have concluded that it is not, and moreover that ID cannot uncouple itself from its creationist, and thus religious, antecedents.
>
> (*Kitzmiller v. Dover*, 2005, p. 136)

Given the evidence for and against ID, the judge concluded that the school board's policy of including it was nothing less than "breathtaking inanity ... resulting [in an] utter waste of monetary and personal resources" (*Kitzmiller v. Dover*, 2005, p. 138).

And what about the disguised "Teach the Controversy" IDC approach that we considered in this article? The Court spoke to that as well, in equally blunt terms:

> Moreover, ID's backers have sought to avoid the scientific scrutiny which we have now determined that it cannot withstand by advocating that the *controversy*, but not ID itself, should be taught in science class. This tactic is at best disingenuous, and at worst a canard.
>
> (*Kitzmiller v. Dover*, 2005, p. 89)

Case closed.

References

AAUP (2005). Teaching evolution: Resolution against teaching intelligent design.
Applegate, D. (2001). Evolution opponents on the offensive in Senate, House. American Geological Institute Government Affairs Program.
Behe, M. (2005). Design for living. *The New York Times*. New York.
Campbell, J. A. and S. C. Meyer (2005). Teach scientific controversy about origins of life. *Chattanooga Times Free Press*. Chattanooga, Tennessee: F1.
Cowan, D. (2005). Teaching students to be "competent jurors" on evolution. *Christian Science Monitor*. May 31, 2005.
DeWolf, D. K. (1999). Teaching the origins controversy: A guide for the perplexed. A Special Discovery Report, Discovery Institute.
Dover Area Board of Directors, 2004). Biology Curriculum Press Release.
Discovery Institute (1999). The Wedge Strategy, Discovery Institute Center for Renewal of Science and Culture.
Discovery Institute (2005). *How to teach the controversy over Darwin legally*. USA: ColdWater Media & Discover Institute.
Edwards v. Aguillard (1987). Edwards, Governor of Louisiana, et al. v. Aguillard et al. Appeal from the United States Court of Appeals for the Fifth Circuit, U.S. Supreme Court. 482 U.S. 578.
Forrest, B. and P. R. Gross (2003). *Creationism's Trojan horse: The wedge of intelligent design*. New York: Oxford University Press.
Johnson, P. E. (1999). Evolution and the curriculum: A conversation with Phillip Johnson and Gregg Easterbrook. *Center Conversations*, EPPC Online.
Johnson, P. E. (2003). *American Family Radio* interview. *Today's Issues*. January 10, 2003.
Kitzmiller v. Dover (2005). Tammy Kitzmiller, et al. v. Dover Area School District, et al. Judge Jones, United States District Court for the Middle District of Pennsylvania.
Krebs, J. and S. Case (1999). Creationists secretly authored Kansas science standards, Kansas Citizens For Science Members Charge. News Release.
Lenski, R. E., Ofria, C., Pennock, R. T., & Adami, C. (2003). The evolutionary origin of complex features. *Nature, 423*, 139–144.
LeVake v. Independent School District #656 et al (2001) *LeVake v. Independent School District* No.656, 625 N.W.2d 502 (Minn. App. 2001), cert. denied, 122 S.Ct. 814 (2002).
Meyer, S. C. (2002). Teach the controversy on origins. *Cincinnati Enquirer*. Cincinnati.
Meyer, S. C. and J. A. Campbell (2004a). Controversy over life's origins: Students should learn to assess competing theories. *San Francisco Chronicle*. San Francisco: B9.
Meyer, S. C. and J. A. Campbell (2004b). Incorporate controversy into the curriculum. *Atlanta Journal Constitution*. Atlanta, GA: Q1, 14.
Meyer, S. C. and J. A. Campbell (2005). Teach the controversy. *The Baltimore Sun*. Baltimore, Maryland: 17A.
Miller, G. (2002). P. D. Eugenie C. Scott, E. Director and I. National Center for Science Education.
NCSE Staff (2002). Analysis of the Discovery Institute's Bibliography. <http://www.ncseweb.org/resources/articles/3878_analysis_of_the_discovery_inst_4_5_2 002.asp>
Peloza v. Capistrano Unified School District (1994) *John E. Peloza v. Capistrano Unified School District, 37 F. 3d 517*.
Pennock, R. T. (1999a). *Tower of Babel: The evidence against the new creationism*. Cambridge, MA: The MIT Press.
Pennock, R. T. (1999b). Of design and deception. *Science & Spirit* (Oct/Nov.).
Pennock, R. T. (2000). Lions and tigers and apes, oh my!: creationism vs. evolution in Kansas. *AAAS Dialogue on Science, Ethics and Religion*.
Pennock, R. T. (2002). Should creationism be taught in the public schools? *Science & Education, 11*(2).

Pennock, R. T. (2004). DNA by design?: Stephen Meyer and the Return of the God Hypothesis. *Debating Design*. M. Ruse and W. Dembski. New York, Cambridge University Press: 130–148.

Princehouse, P. (2006). Teaching Biology in Ohio: With God, all things are possible* (*except macroevolution). R. T. Pennock & Susan Spath (Ed.) *Defending Evolution Education: A Guidebook for Citizens for Science*.

Richardson, M. K., J. Hanken, et al. (1998). Haeckel, embryos, and evolution. *Science* 280(5366): 983.

Stephens, S. (2002). Intelligent design advocate lauds state plan on teaching evolution. *The Plain Dealer*. Cleveland.

Sulon, B. (2005). Institute didn't favor policy, transcript says. *The Patriot News*. Harrisburgh.

Webster v. New Lenox School District (1990) *Webster v. New Lenox School District #122*, 917 F. 2d 1004.

Williams, R. (2003). The Steve Project, The Science Show, *ABC.science*.

6

The Scientific Enterprise and Teaching about Creation

Michael Poole

Scientific accounts of 'origins' often give rise to theological questions about 'Creation-by-God'. If scientific accounts of origins appear to contradict religious ones, difficulties may arise in science education for religious science students. This is especially so for those who in their faith communities are taught what is loosely termed *creationism*—young-Earth creationism to be precise, although there are other sorts. Finding that this entails the wholesale rejection of substantial blocks of orthodox (Big Bang) cosmology, geology, physics and biology, the student may see the only alternatives to be a loss of faith, maintaining a 'Sunday mind' and a 'weekday mind', or the rejection of conventional science. In this chapter I seek to identify a number of related difficulties, to comment upon them and then to make some educational recommendations. I shall argue that some rapprochement is possible between the parties at variance, while recognizing that a key issue, the age of the Earth, obviously cannot be resolved by both parties agreeing to move some way towards each other's position! I shall argue that teaching about Creation and a professional dedication to the scientific enterprise can, with integrity, go hand in hand. How this might be so requires an understanding of what some see, on religious grounds, as *areas of difficulty*. Before identifying these areas, it is useful to understand the distinction between *Creation* and *creationism*, as well as to be aware of the dramatic resurgence of early views on *flood geology*. Having said this, it is perfectly possible, at a first reading, to go straight to the heading, *areas of difficulty,* and begin there.

A disturbing consequence of the prevalence of young-Earth views among certain religious communities has been documented by Francis *et al.* (1990) in studies of teenagers' beliefs. They conclude one such study, of Scottish teenagers (p. 16), by saying that the results

> ... confirm the findings ... among sixteen-eighteen year olds that both scientism and the perception of Christianity as necessarily involving creationism [defined as 'the view that the accounts of the origins of creation in Genesis are literally true and that evolutionary theories are false'] are important factors in helping to shape both attitudes towards Christianity and interest in science. In particular, both factors contribute to making it more difficult for pupils to combine a positive attitude towards Christianity with a high level of interest in science. If it is thought desirable that pupils should be enabled to develop both positive attitudes towards Christianity and interest in science, there need to be opportunities within the religious education curriculum both to exam-

ine the nature, scope and limitations of scientific enquiry and to recognize that there is a range of acceptable views within Christianity itself on the authority of scripture and the nature of origins.

In England, the religious education curriculum is, by statute, the responsibility of 151 or so Local Education Authorities, through the production and administration of *Agreed Syllabuses*. A study (Bausor & Poole, 2002, 2003) of the science-and-religion content of these syllabuses indicates that, of over a hundred of these syllabuses that were searched, about two thirds of them contained material relating to issues of science-and-religion. Three significant omissions were identified: (i) a comparison of how *language* is used in science and in religion (ii) the *nature of explanations* in science and in religion (iii) the history of the *'conflict thesis'*.

The introduction of a Non-Statutory National Framework for Religious Education (QCA, 2004) brought with it substantial recommendations for teaching about the relationships between science and religion. But attitudes to religious education and its provision—or lack of it—vary greatly in different countries and my concern here is with issues raised by beliefs in a young Earth.

'Creation', 'Creationism' and Flood Geology

Creation
'Creation', in its theological sense, refers to the 'bringing-into-being-of-everything-by-God' and the consequent relationship between Creator and creation. Some theologians also include 'sustaining' within the concept of creation. Others treat the ideas of *creating* and *sustaining* as distinct though related concepts, as do the New Testament writers of Hebrews 1:1-3 and Colossians 1: 16f. I shall follow their practice.

It is necessary to distinguish between (1) the religious concept of *Creation* (2) the word *creation* as borrowed from theology to describe the 'bringing-into-being' of new works of art, ideas, fashions in clothes, etc. and (3) scientific accounts of *origins*—of the universe, the Earth, life on Earth and humankind. The theological concept of Creation is independent of any particular mechanisms (e.g., stellar, chemical and organic evolution) by which matter came into being. The claim that it is necessary to choose between 'Creation' and 'evolution' involves some kind of *category mistake*. To illustrate the point through a comparison, the 'creation' of a new type of car is independent of whether the cars are assembled by hand or by automation. But it would sound very odd to say that a new type of car had not been created because 'automation *did* it'. Yet claims like 'humans weren't created by God; evolution *did* it' tend to escape critical scrutiny.

Creationism
It can be seen from the religious concept of *Creation* that all Jews, Christians and Moslems of conventional faith are 'creationists' in the strict sense that they

believe that God created everything. A problem arises because the word 'creationist' has misleadingly been annexed by particular groups, who also claim to specify time-scales and processes. The resulting confusion leaves adherents to the three Abrahamic religions without a word to identify, unambiguously, their belief in God as Creator.

The issues to be discussed in this chapter are of particular importance to Christians and Moslems, but apparently less so to practicing Jews. Since a chapter of this book is devoted to a Moslem perspective on creationism, I shall concentrate on offering a Christian perspective.

By the time Darwin's *Origin* was published in 1859, many Christians had already accepted the idea of an Earth of great antiquity. Two prevalent interpretations of Genesis 1 were taken to justify compatibility between the geological and biblical accounts. One was the *gap* theory which postulated an unspecified time interval between the first two verses, during which some catastrophic event was believed to have taken place. The subsequent verses were seen as describing a re-creation or reconstruction after the supposed catastrophe. The other interpretation was the *day/age* theory which identified the 'days' of Creation with geological ages. Both these interpretations of the text are now seen as questionable. A variant of these old-Earth creationism positions was that of progressive creationism. This viewed God as having created many species which subsequently multiplied by mutation and selection. The essence of all these positions is that something additional is needed to the forces with which nature has already been endowed by God.

Flood Geology

The Seventh Day Adventists were founded in 1863, their leader being Mrs. Ellen White. She was believed to have a gift of prophecy and her visions were viewed as comparable in authority to the Bible. The Seventh Day Adventist observance of Saturday as the Sabbath was integrally linked with belief in a literal creation week of 6 x 24h days. Mrs. White espoused a view of the Genesis flood as having covered the whole globe, a view that had lost favor among geologists some decades earlier. Charles Lyell's (1830–3) *Principles of Geology* had brought to an end the widespread perception of Noah's Flood as a major geological agent. Another Seventh Day Adventist, George McCready Price, a schoolteacher, found Ellen White's position persuasive. At the beginning of the twentieth century he began a prolific period of publicizing such views (Numbers, 1992, pp. 73ff, 299).

A landmark in this resurgence of flood geology came with the (1961) publication of *The Genesis Flood* (1961) by Henry Morris, a Professor of Hydraulic Engineering and John Whitcomb Jr, a Professor of Old Testament. It was a hard back, glossy-paper book of over 500 pages, copiously furnished with footnotes, and it was to have a huge influence among conservative evangelicals on both sides of the Atlantic and beyond. For our present concern it is significant in the way it shifted populist belief in an ancient Earth, allegedly supported by the Gap and the Day/Age theories, to a geologically young Earth of not more than 10 000 years. It was the extent of this turnabout that resulted in the term 'creationism' becoming synonymous with *young-Earth creationism* — from about 1980 onwards.

Areas of Difficulty
Seven main religious grounds for concern about the portrayal of science are listed below. The first is unique to young-Earth creationists. The following six have been expressed by both young-Earth creationists and religious believers who do *not* accept a young Earth.

1. Age of the Earth
The early chapters of the book of Genesis are thought by young-Earth creationists to teach a relatively young Earth of some 10,000 years in age compared to the received view of about 4.6 billion years in a universe of some 14 billion years.

This interpretation of Genesis is at variance with current cosmology and with the life and earth sciences as well as with relevant branches of physics. Followers of young-Earth teaching reject the Big Bang and with it the religiously interesting Anthropic Cosmological Principle. This principle is sometimes dubbed the 'Goldilocks Effect' since the physical constants of nature, like Baby Bear's bed and porridge, are 'just right' for life as we know it to have arisen. In some cases the precision is accurate to one part in 10^{60}—something that gives rise to one of those questions raised by science, which science cannot answer, namely, 'Is there purpose or plan behind the universe?' This isn't a question that can be answered by pointing out that the *inflationary theory* of the early universe may result in this apparent 'fine tuning' of the physical constants. That simply pushes back the question to 'Why were features of the early universe such as to give rise to an inflationary universe which in turn …?' Another interpretation offered is that perhaps we are just one universe 'within' a multiverse, one that happens to be right for life among the myriads that are not. Semantic difficulties loom over the meaning of *universe*. Violations of Ockham's razor, that 'it is vain to do with more what can be done with fewer', abound. On present thinking, communication with any 'alternative universes' would not be possible, making testing out of the question and leaving the idea as a piece of speculative metaphysics.

It has to be said that there are huge amounts of evidence in favor of an ancient Earth. People not familiar with the relevant disciplines of cosmology, physics, geology and biology often seem unaware of the coherence of views about the vast ages of the Universe and the Earth arising from studies in each of these major sciences. It should nevertheless be remembered that religious believers who have the same high regard for the early chapters of Genesis as do young-Earth creationists are among those academics who regard as conclusive the evidence for an ancient Earth. It needs to be asked of those who hold young-Earth views whether their particular interpretation of the early chapters of Genesis might be incorrect. After all, the value of science in guarding against poor exegesis has long been recognized, as it was in Galileo's day over claims about the Sun's movement around the Earth based on Psalm 19. A major factor in this issue is surely the need to recognize distinctive *literary genres* in the biblical writings. Not all Bible passages employ the same ones. Over thirty different *genres* have been identified. Both in science and in religion it is important to be aware that there are other forms of language than the literal. Each discipline resorts to the 'it-is-as-if' language of metaphors and models when trying to express that which is novel, invisible and conceptually difficult. This recognition does not downgrade the Bible but treats it much more carefully. In addition to paying attention to *literary genre* the reader needs to give careful attention to other key matters such as searching out the original context, the intention of the author and the meaning of the writings for today.

It should not be thought, however, that figurative interpretations arose simply as a way of getting around geological problems about the antiquity of the Earth. Long before the age of the Earth or the processes of evolution were investigated by modern science, early Church Fathers such as Origen and Augustine had commented on matters of the literary *genre* of these chapters, arguing for a figurative understanding. Origen asked, in the third century, 'What man of intelligence, I ask, will consider as a reasonable statement that the first and second and the third day, in which there are said to be both morning and evening, existed without sun and moon and stars...?'. (Lucas, 2001, p. 95). It could be argued from a straight reading of Genesis 1 that the concepts of solar or sidereal (star-based) days had no meaning prior to 'day' four. Augustine (354–430 AD), in his work *The Literal Meaning of Genesis,* was even more forthright in his comments and his phraseology would not be counted as politically correct today! In one chapter entitled, 'On interpreting the mind of the sacred writer. Christians should not talk nonsense to unbelievers', he wrote:

> Usually, even a non-Christian knows something about the earth, the heavens, and the other elements of this world ... and this knowledge he holds to as being certain from reason and experience. Now, it is a disgraceful and dangerous thing for an infidel to hear a Christian, presumably giving the meaning of Holy Scripture, talking nonsense on these topics; and we should take all means to prevent such an embarrassing situation, in which people show up vast ignorance in a Christian and laugh it to scorn ... If they find a Christian mistaken in a field which they themselves know well and hear him maintaining his foolish opinions about our books, how are they going to believe those books in matters concerning the resurrection of the dead, the hope of eternal life, and the kingdom of heaven ... (Augustine, c. 401–15).\

Many arguments have been advanced by young-Earth creationists for an Earth aged some 10,000 years. They include 'thickness of Moon dust', 'shrinking Sun', 'decay of the Earth's magnetic field', 'cooling of the Earth', 'change in the velocity of light', 'salinity of the oceans', 'coal and oil being formed quickly', and so on. Such arguments tend to be restated again and again in successive publications even when it is made known that they do not withstand scrutiny by academic scientists *in the relevant fields*. Collectively, the number of supposed 'arguments' does not add to their weight if each is deficient.

Educationally, some have argued that a young-Earth account should be presented to pupils as a viable 'alternative view' in the name of 'fairness'. This, however, would be inappropriate in science lessons since mainstream science finds the case for a young Earth no more viable than the 'alternative views' of an Earth-centered planetary system or a phlogiston theory for redox reactions. Such views might be referred to *en passant* as views that were once held in the history of science, but reasons should be given why orthodox science no longer considers them realistic.

Large sums of money have been invested, particularly in the US, in promoting young-Earth views and opposing evolution—other than microevolution. The increasing complexity of the literature limits the ability of the average churchgoer to evaluate its quality. Hence there is the danger of a superficial and ungrounded acceptance—'Although I don't understand this, they seem to know what they are talking about'. Whole congregations in some churches, under the influence of a minister, a member of the congregation or a denominational newspaper, seem to regard the current vogue for a young-Earth as orthodoxy. It hasn't always been that way, although with tendentious book titles like 'Evolution or Creation', 'Creation vs. Evolution' and 'The Truth: God or Evolution', little choice seems to be offered to the faithful.

2. Chance

The place of chance/random variations in natural selection is used to imply an accidental, rather than a designed universe.

The concepts of 'chance' and 'randomness' are complex and their theological significance has attracted considerable attention (Bartholomew, 1984). The terms may indicate, for example, 'unpredictable from prior data' or 'the intersection of two causally unrelated events'. In popular, non-technical parlance, the term 'chance' is often used—and here lies a problem—as a synonym for 'accidental' or 'unplanned'. It cannot, however, be inferred that a total system, of which chance/random events form a part, is itself 'accidental' or purposeless. Evolution by natural selection can after all be seen as an ingenious way of ensuring that available ecological niches continue to be filled, provided food supplies and environmental features do not change too rapidly for adaptation to take place. In the evolutionary picture, the concept of design was radically altered from William Paley's formulation, but did not disappear. Darwin, although he confessed himself "bewildered", wrote on one occasion to Asa Gray, 'I can see no reason why a man, or other animal, may not have been aboriginally produced by other laws, and that all these laws may have been expressly designed by an omniscient Creator, who foresaw every future event and consequence.' (Brooke, 1985, p. 56)

Frederick Temple (later Archbishop of Canterbury), who preached the official sermon for the (1860) British Association for the Advancement of Science meeting the day after the legendary Wilberforce-Huxley 'debate', subsequently wrote that 'What is touched by this doctrine [of evolution] is not the evidence of design but the mode in which the design was executed ... In the one case the Creator made the animals at once such as they now are; in the other case He impressed on certain particles of matter ... such inherent powers that in the ordinary course of time living creatures such as the present were developed ... He did not make the things, we may say; no, but He made them make themselves' (Temple, 1885, pp. 114f). Charles Kingsley took up this idea and was quoted in favor by Darwin in *The Origin of Species*.[1]

If it is thought that a process of *chance/random changes + selection* contradicts the idea of an Intelligent Being, it is of note that what has been termed 'Darwinian design' has been used for aerofoil design (Poole, 1994, 52f). *Evolution Strategy*, as it is properly known, has also been used, among other projects, for the optimization of a two-phase (fluid-steam) jet nozzle (Rechenberg, 1989, pp. 116f) This work of Rechenberg, published almost two decades ago, utilized a simple mutated offspring model. At about the same time the sophisticated tool of *genetic algorithms* was invented by Holland and has been extensively developed by him and others since. It arose out of investigating how adaptation works in nature and how the processes involved might be used as an analogy in computing systems to seek out solutions to complex combinatorial problems that would otherwise require unrealistic amounts of computing time. Typically, genetic algorithms can be helpful where a problem has too many possible solutions (even for a computer) to evaluate all of them in a realistic time, but where it is possible to ascertain in a short time the effectiveness of a particular solution.

The process is clearly described by Mitchell (1998, pp. 1–16) in *An Introduction to Genetic Algorithms*. Briefly, *recombination* of gametes during sexual reproduction is mimicked by a process called *crossover*. '*Chromosomes*' are analogies for possible solutions to problems, represented most simply by strings of binary digits (*bits*—0 or 1) or by more complex forms such as binary trees or *if-then* rules. Mutations are imitated by changing a *bit* or *bits* at randomly selected locations. Initial *selection* is from the 'fittest chromosomes', i.e., those that are better solutions to the problem. The process is iterative and, combined with parallel processing, provides a powerful tool for searching out huge numbers of potential solutions to problems. The technique is used by evolutionary biologists, social scientists, engineers and others. Among its strengths is its applicability to a wide range of optimization problems. It hardly needs pointing out that if such biological analogues can be used by intelligent human agents to achieve desirable ends, it can scarcely be claimed that evolutionary processes are necessarily inimical to divine purpose and plan in nature. Evolution by natural selection does not eliminate the possibility of design; it changes its form.

---

[1] In the sixth and last edition the quotation can be found on p. 658 [London, John Murray].

I have already referred to the Anthropic Cosmological Principle as a candidate for a contemporary argument-from-design. A more recent variant of the argument-from-design has arisen from the Intelligent Design (ID) Movement. Its adherents come from a wider theological base than is the case with young-Earth creationism, and their basic argument is that there are in nature certain phenomena which are so 'irreducibly complex' that they must have been the products of Intelligent Design and cannot be capable of a 'natural' explanation. Some of the arguments are mathematically complex and the literature is multiplying rapidly. One difficulty with the idea is a theological one. Although the Bible teaches that the world is planned by God, it also teaches that 'the whole creation has been groaning as in the pains of childbirth right up to the present time' (Romans 8:22 NIV). Human sins of commission and omission (with respect to the creation mandate to take on managerial responsibility for the Earth) account for some of this; so any search for design encounters mixed messages. Another difficulty is a philosophical one arising out of the slipperiness of the notion of 'irreducible complexity'. To claim 'irreducible complexity' on account of the current lack of any scientific explanation for a biological phenomenon is a risky business. A biological explanation may be found tomorrow. History is littered with failed claims of this kind and, despite the denials of its followers, ID appears to be a contemporary example of the counterproductive strategy, referred to later in Section 6, known as the 'God-of-the-gaps'.

As I see it, the Bible portrays God as the creator and the sustainer moment-by-moment of everything there is. It does not picture him as the God of semi-deism who pokes the tweezers in from time to time to tweak something that he hasn't endowed with the capacity to function towards his chosen ends. As Charles Kingsley expressed it in his approval of Darwin's theory, 'they find that now they have got rid of an interfering God—a master-magician, as I call it—they have to choose between the absolute empire of accident, and a living immanent, ever-working God.' (Kingsley, F., 1887, p. 171) After all, as one clergyman (A. L. Moore, 1843–1890) put it, 'a theory of occasional intervention implies as its correlative a theory of ordinary absence' (cited in Moore, J. R., 1979, pp. 263f). Howard Van Till has written extensively on this matter of the 'functional integrity of creation' and more recently has developed it in greater detail in what he calls the *Robust Formational Economy Principle (RFEP)*. In a universe that satisfies this principle: 'Nothing would be missing from the universe's resources, capabilities, or potentialities that would prevent it from actualizing ... any type of physical structure ... or any type of organism that has appeared in the course of time.' (Van Till, 2002, p. 232) This belief underpins the practice of scientists of all faiths and of none and, Van Till argues, is entirely consistent with biblical teaching. ID arguments, by contrast, provide a contemporary version of 'episodic creationism'.

3. Implications of atheism
Science is often presented as an atheistic activity that makes no place for God.

Tensions have been caused in the past by bad science teaching and misleading public presentations of the nature of science as inherently atheistic. There is some truth in the creationist claim that 'what is served up as Science for popular consumption is frequently riddled with subtle atheistic propaganda' (Layfield, 2000, pp. 22f). Science *has* at times—and none-too-subtly—been illegitimately portrayed as an atheistic activity, something that flies in the face of historical studies of the religious beliefs of scientists, past and present. The Golden Age of Islamic science (ninth and tenth centuries) and the Christian *milieu* into which modern (seventeenth century onwards) Western science was born and nurtured bear testimony to the positive contribution religious belief has made to the development of science.

In dissociating science and science education from illegitimate entailments of atheism, students need to be taught sufficient history and philosophy of science to understand the issues involved. In England, for example, it is a statutory requirement that science, in common with other subjects on the curriculum of maintained schools, should promote the spiritual, as well as other aspects of the development of pupils. The category 'spiritual' includes the 'religious', indicating the appropriateness of considerations of the subject matter of this chapter, particularly for 14–16 year olds (Key Stage 4) onwards. For an earlier (1995), but not dissimilar version of *Science in the National Curriculum*, I have indicated places where such issues might naturally occur (Poole, 1998). The most recent version of the Key Stage 4 programme of study (QCA, 2004, p. 37) in science takes the trouble to point out 'that there are some questions ... that science cannot address.'

In the US, two researchers suggested that 'some of the critics of evolution ... have adopted a hyperbolic, aggressive rhetoric suggesting that American educators are engaged in some sort of gigantic conspiracy to undermine traditional religion' (Johnson & Giberson, 2002, p. 248). This prompted them 'to investigate the rôle that evolution plays in the curriculum of the Quincy, Massachusetts, public school system' (*ibid.*, p. 242). They examined curriculum materials used in elementary, middle and high schools as well as interviewing teachers. They found a sensitivity of approach and concluded that 'If, as our research suggests, this strident claim simply is not true, then it would appear that the conservative critics of evolution are fighting an imaginary foe. This is unfortunate' (*ibid.*, p. 248). There have, of course, been places like the former USSR, where atheism has been promoted in science education, and there are still countries where the ideologies seek to foster a similar view of science.

As part of the historical study of the interplay between science and religion, students could usefully consider how the idea of a 'conflict' between science and religion arose and how recent research has altered perspectives. Studies of the *conflict thesis* by historians of science such as professors Bowler, Cantor, Brooke and Russell have resulted in such comments as: 'The old model of inevitable conflict (still visible in the writings of extremists on either side) has been heavily qualified, if not abandoned' (Bowler, 2001, p. 5). Cantor (1991, pp. 290f) summarizes a similar point by saying that '... the conflict thesis is like a great blunderbuss which obliterates the fine texture of history and sets science and religion in necessary and irrevocable opposition. Much historical research has invalidated the conflict thesis.' Brooke (1990, p. 198) refers to 'what nineteenth-century rationalist historians liked to call the warfare between science and religion ... that popular but simplistic formula.' Russell (1989, p. 5) concludes 'that to portray them as persistently in conflict is not only historically inaccurate, but actually a caricature so grotesque that what needs to be explained is how it could possibly have achieved any degree of respectability.' In time, as a result of such studies, popular perceptions may gradually shift. There is evidence of some recent change in media presentations of the topic in the UK.

The 'makes no place for God' clause in the sub-heading above is not sinister, as might appear. Science only deals with the natural world, so talk-about-God is outside of its sphere of competence. Although Newton could write in 1692, 'When I wrote my Treatise [*Principia*] about our System, I had an Eye upon such Principles as might work with considering Men, for the belief of a Deity' (Goodman, 1973, p. 131), it is a tacit agreement nowadays that science does not make reference to First Causes. This enables those of different faiths and of none to work together on a common enterprise. It need be no more surprising to the religious scientist not to find God mentioned in science texts than to find that Henry Ford is not mentioned in the instruction booklet of that make of car.

4. Naturalism
The physical world is presented as all that there is.
The charge of a commitment to *naturalism* necessitates a brief explication of this word since it has different meanings in different contexts. In the present context, a consultation of five dictionaries of philosophy indicates a consensus that *naturalism* accords with the view that finally nothing defies explanation by the methods of the natural sciences. Such a claim goes far beyond the methodological commitment of science to studying the natural world and harks back to Bertrand Russell's imperialistic claim that 'Whatever knowledge is attainable, must be attained by scientific methods; and what science cannot discover, mankind cannot know.' This brief statement epitomizes earlier, but now discredited, attempts of the movement known as *logical positivism* to set up science as the final arbiter of everything that could rationally be believed. So deeply entrenched in society did this view become that for many years it was common to hear the demand 'prove to me *scientifically* that God exists'. But why *scientifically*? Religious questions such as 'Is there something (God?) other than the natural world to which the natural world owes its existence?' cannot be answered by science, the study of the natural world. *Naturalism* is a metaphysical

belief held by some people, including some scientists but is not entailed by science itself. Indeed, as illustrated earlier, science prompts some questions that science itself cannot answer. Another of these (metaphysical) questions is 'Why is there something rather than nothing?'

5. Explanation
Scientific explanations are frequently portrayed as ruling out religious ones.
Scientific explanations of mechanisms are logically compatible with religious explanations of divine agency and purpose and should not be presented as alternatives (Poole, 2002). Failing to distinguish between these two different types of reason-giving explanations is an example of a *type-error* in explanation. So widespread is this particular type-error that it has a name of its own, the 'God-of-the-gaps'. Its origin lay in a sense of threat felt by some religious believers that science was pushing God out of the picture. So they pointed to what science had not yet explained and said 'that's God's doing' as though the rest was not! As an apologetic device it was counterproductive since its consequence was that the more science explained, the less room there was for God. This resulted in a fear of science by some believers and increased enthusiasm for science by some non-believers. But since Judaism, Christianity and Islam portray the known as well as the unknown as God's activity, the 'God-of-the-gaps' stratagem is theologically unsound.

6. Reification
Evolution by natural selection is portrayed as replacing what was formerly attributed to God.
Many have treated evolution as a kind of surrogate religion (Midgley, 1985). Extending the theory beyond the biology to include metaphysics is commonly referred to as *evolutionism,* a sub-section of *scientism.* While the activity of a creator God is denied, concepts such as 'nature', 'evolution', 'natural selection', and 'chance' are vested with the powers formerly attributed to God and credited with the ability to *do* things like 'create', 'make', 'choose' and 'build'! If this is simply a sloppy way of talking, little harm is done, except perhaps causing some confusion. But if it is intended seriously to make an assertion like 'God didn't make living things; evolution *did* it', it earns the title of *the fallacy of reification*—confusing a concept with a real object or cause. Careless talk of this kind creates unnecessary problems for religious believers and needs to be avoided in class.

7. Evolutionism
Extrapolations of evolutionary biology to non-biological areas such as 'evolutionary ethics' are used to present a very different moral stance to a Christian position.
Durant (1985, p. 34) supports this point, saying:

> ... much of the energy of the creationist movement arises from a sense of moral outrage at the advance of an evolution-centred world-view

that has the audacity to parade its secular, liberal values as if they were the objective findings of science. Here at least, if not in matters of biological fact and theory, creationism has a point of which the scientific community might do well to take heed.

Attempts to extend the biological theory of evolution to embody metaphysical ideas have already been referred to. On this matter Durant (*ibid.,* p. 33) comments that:

> In the past, attempts to derive optimistic lessons from biology concerning the future of humankind have owed far more to prior religious or political convictions than they have to any independent insights derived from science; and as the case of Julian Huxley illustrates, this has been the case even where those involved have been major authorities on Darwinism. There is nothing in a scientific training, it would seem, that immunizes a person against their own prejudices.

Flew (1970, p. 30) commented about Huxley's attempt to derive ethical values from biology, saying:

> There is, surely, something very odd, indeed pathetic, in Huxley's attempt to find in evolutionary biology 'something, not ourselves, which makes for righteousness'. For this quest is for him a search for something, not God, which does duty for Divine Providence. Yet if there really is no Divine Providence operating in the universe, then indeed there is none; and we cannot reasonably expect to find in the Godless workings of impersonal things those comfortable supports which—however mistakenly—believers usually think themselves entitled to derive from their theistic beliefs.

Although the three quotations in this section were penned some time ago, the points they make retain a contemporary ring. The debates about attempts to derive 'ought' from 'is' continue; and it is salutary to reflect that the three mutually incompatible ideologies of Victorian Capitalism, Communism and Nazism all claimed to justify their beliefs from evolutionary theory.[2]

Finally, in drawing together the points already made, I offer the following recommendations for the teaching of science:

Recommendations

Pupils should be taught, at appropriate places within science education (and, where applicable, religious education) that:

> 1. Although individual scientists may have various religious beliefs or none, there is nothing inherently atheistic about the scientific enterprise.

---

[2] This section [7] draws upon my (1995) *Beliefs and Values in Science Education,* p. 128, Buckingham: Open University Press, which gives a more detailed treatment of some of the issues addressed in this chapter.

2. Science studies are limited to the natural world but it does not follow that therefore there is nothing other than the natural world.
3. Scientific explanations of the mechanisms of the world are logically compatible with religious explanations of divine agency and purpose.
4. A distinction needs to be made between the religious doctrine of creation, held by Jews, Christians and Moslems, as the 'bringing-into-being' of everything by God, and varieties of 'creation*ism*' which seek, additionally, to specify particular mechanisms and time scales.
5. Over-extravagant claims for science and for evolution need to be guarded against.
6. There is a need to distinguish between historical and folklore accounts of such episodes as the Galileo affair and the Wilberforce-Huxley encounter.
7. Belief in an Earth about 10,000 years old is incompatible with the current, well-attested scientific view of some 4.6 billion years, in a universe of 14 billion years; and would be inappropriate to be taught as a viable 'alternative view' in science lessons.
8. Both science and religion use metaphors and models to refer to what is *new, invisible,* and *conceptually difficult.*
9. The context of writings and the original intentions of authors are essential aspects of understanding ancient texts like the Bible, as is recognizing the literary forms or *genres* of individual passages.
10. The reification—confusing a concept with a real object or cause—of 'nature', 'evolution', 'natural selection', 'DNA', 'the laws of nature', 'time', 'chance' and the like should be avoided so as not to appear to invest them with the ability of sentient beings to 'choose', 'build', 'manufacture', 'create', 'search', 'make', etc.

References

Augustine (1982). *The Literal Meaning of Genesis*, Vol. 1, (trns. J. H. Taylor), New York: Paulist Press [commenced by Augustine AD 401].

Bartholomew, D. J. (1984). *God of Chance,* London: SCM Press.

Bausor, J. & Poole, M. W. (2002). Science-and-religion in the agreed syllabuses—An investigation and some suggestions, *British Journal of Religious Education, 25*(1), 18–32.

Bausor, J. & Poole, M. W. (2003). Science education and religious education: possible links? *School Science Review, 85*(311), 117–24.

Bowler, P. (2001). *Reconciling Science and Religion: The Debate in Early-Twentieth-Century Britain*, Chicago: University of Chicago Press.

Brooke, J. H. (1985). The Relations between Darwin's science and his religion in *Darwinism and Divinity,* Durant, J. (ed.), Oxford: Blackwell.

Brooke, J. H. (1990). The Galileo affair: teaching AT 17, *Physics Education, 25*(4), 197–201.

Cantor, G. (1991). *Michael Faraday: Sandemanian and Scientist,* Basingstoke: Macmillan.

Durant, J. (ed.) (1985). *Darwinism and Divinity,* Oxford: Blackwell.

Flew, A. G. N. (1970). *Evolutionary Ethics,* London: Macmillan.

Francis, L. J., Gibson, H. M. and Fulljames, P. (1990). Attitude towards Christianity, Creationism, scientism and interest in science among 11–15 year olds, *British Journal of Religious Education, 13*(1), 4 -17.

Goodman, D. C. (1973). *Science and Religious Belief 1600–1900: A Selection of Primary Sources,* p. 131, Letter 1 (10 December 1692) of four letters from Sir Isaac Newton to Dr Bentley concerning arguments in Proof of a Deity, published London 1756, Dorchester: John Wright/Open University.

Johnson, T. R. & Giberson, K. (2002). The teaching of evolution in the public school: A case study analysis. *Perspectives on Science and Christian Faith, Journal of the American Scientific Affiliation, 54*(4), 242–248.

Kingsley, F. (1877). *Charles Kingsley, His Letters and memories of his Life,* Vol.2, London, cited in Meadows, A. J. (1975). Kingsley's Attitude to Science. *Theology,* LXXVIII, No 655.

Layfield, S. (2000). *The Teaching of Science: A Biblical Perspective,* a lecture given at Emmanuel College, Gateshead on 21 September 2000, Newcastle upon Tyne: The Christian Institute.
Lucas, E. (2001). *Can We Believe Genesis Today?*, (2nd Ed), Leicester: Inter-varsity Press.
Midgley, M. (1985). *Evolution as a Religion: Strange hopes and stranger fears,* London: Methuen.
Mitchell, M. (1998). *An Introduction to Genetic Algorithms,* London: MIT Press.
Moore, J. R. (1979). *The Post-Darwinian Controversies.* Cambridge: Cambridge University Press.
Numbers, R. L. (1992). *The Creationists,* New York, Alfred Knopf.
Poole, M. W. (1994). A critique of aspects of the philosophy and theology of Richard Dawkins. *Science and Christian Belief, 6*(1), 52f.
Poole, M. W. (1998). *Teaching about Science and Religion – Opportunities within Science in the National Curriculum.* Abingdon: Culham College Institute.
Poole, M. W. (2002). Explaining or explaining away?—The concept of explanation in the science-theology debate. *Science and Christian Belief, 14*(2), 123–142.
Qualifications and Curriculum Authority (2004). *The Non-Statutory National Framework for Religious Education.* London: QCA.
Rechenberg, I. (1989). Evolution strategy: Nature's way of optimization. In Bergmann, H. W. (ed.) *Optimization: Methods and applications, possibilities and limitations,* pp 106–126, Berlin: Springer-Verlag, Berlin.
Russell, B. (1970-2). *Religion and Science.* Oxford: Oxford University Press.
Russell, C. (1989). The conflict metaphor and its social origins. *Science and Christian Belief, 1* (1), 3–26.
Science: The National Curriculum for England, (2006). London: Department for Education and Employment/ Qualifications and Curriculum Authority.
Temple, F. (1885). *The Relations Between Religion and Science.* The Bampton Lectures for 1884, London: Macmillan.
Van Till, H. J. (2002). Is the creation a "right stuff" universe? *Perspectives on Science and Christian Faith, Journal of the American Scientific Affiliation, 54*(4), 232–239.

The Theory of Evolution: Teaching the Whole Truth

Shaikh Abdul Mabud

Evolution—A Fact or Theory?

The world-renowned evolutionist S. J. Gould wrote:

> Well, evolution is a theory. It is also a fact ... Facts are the world's data. Theories are structures of ideas that explain and interpret facts. Facts don't go away when scientists debate rival theories to explain them. Einstein's theory of gravitation replaced Newton's in this century, but apples didn't suspend themselves in midair, pending the outcome. And humans evolved from ape-like ancestors whether they did so by Darwin's proposed mechanism or by some other yet to be discovered.
>
> (Gould, 1984, p. 118)

Gould would like us to believe that evolution is as true as apples falling to the ground. For him, there is no controversy about evolution. The contention is whether apples fall as a result of a mischievous lad throwing stones at them or due to some hidden force that Newton or Einstein thought they had discovered. Gould wants to tell us that, as the falling of an apple is an undisputed reality, so is evolution an established fact in the scientific community. This distinction between fact and theory has been maintained by evolutionists from the days of Darwin. Listening to such authoritative statements from renowned scientists, the nonscientific community would be inclined to believe that scientists must have seen evolution happening in nature. This would mean that there is no doubt that our species has evolved, even though the exact mechanism is a matter of debate, which in itself is a healthy sign of the progress of science and the human quest for the truth.

It is natural to ask why scientists think that evolution has occurred and is still occurring, and what is the basis of this conclusion? The only reliable answer that I have found from the literature is that, it is the consensus of the scientific community that confirms the happening of evolution. Michael Ruse, an expert witness in the *McLean vs. Arkansas* court case (Ruse, 1988), defending the case for evolutionism, maintained the distinction between the happening of evolution (fact) and how evolution happened (theory). When he said that he did not know of any credible scientist who questioned that evolution had actually happened, he was asked to explain why scientists agree on evolution. What follows is an excerpt of the conversation that went on in the court:

Q: You say that scientists agree that evolution happened. Why is that?
A: Because the evidence is absolutely overwhelming. It convinces the unbiased observer beyond any reasonable doubt.
...
Q: Do scientists generally agree now about how evolution happened?
A: No, not at all. With respect to this issue of how evolution happened there is still much debate.

(Ruse, 1988, pp. 288–289)

This view of "overwhelming" evidence providing the most "reasonable" proof of evolution has been followed in the teaching of biology in schools in almost all the countries of the world. The assumption is that the case in favor of evolution having occurred is not seriously threatened and that every new piece of data from every relevant field of science confirms that evolution has happened. If this assumption is true, we are perfectly justified in teaching our students evolution as an established scientific fact, but if it is not, we must present a balanced view of evolution, that is, arguments both for and against the theory of evolution.

Although Darwin is famous for his theory of evolution, the idea that new species can arise out of old ones had been around for thousands of years before the publication of his book, *The Origin of Species*, in 1859. Then it was not known as "evolution" but as "development" or "descent with modification" or by other names. Regardless of what they term this process, the ultimate conclusion, that evolution happened, is an indisputable fact amongst evolutionists. However, Darwin himself never claimed to have provided direct evidence of evolution or of the origin of species; what he did claim was that if evolution has occurred, a number of facts, otherwise inexplicable, can be easily explained.

Teaching Biological Evolution in Schools

The intention of this chapter is to look at the way evolution is presented in some biology textbooks used in British schools. It is not my primary purpose to present here an argument in support either of the theory of evolution or of its critics, as that would require much more comprehensive and rigorous analyses of various issues involved. Instead, I would like to show that the presentation of the existence of various forms of life in school textbooks and lessons is highly biased in favor of the theory of evolution and the scientific findings contrary to the evolutionary perspective are hardly mentioned. Indeed, the teaching of evolutionary biology in school is characterized by a deplorable lack of a balanced approach. What I have set out to do in this chapter is to argue that there are deep flaws in the way that the theory of evolution is taught in schools, and to give an idea of how I think the present situation could be remedied.

As we have seen above, Michael Ruse says that "the evidence is absolutely overwhelming." Of course, there is no denying the fact that there is evidence, but the crucial question is what that evidence is. The most striking evidence has come from the fossil record, as is indicated by the following quotation:

> For that matter, what better transitional form could we desire than the oldest human, *Australopithecus afarensis*, with its ape-like palate, its human upright stance and a cranial capacity larger than any ape's of the same body size, but a full 1000 cc below ours. If God made each of the half dozen human species discovered in ancient rocks, why did he create in an unbroken temporal sequence of progressively modern features—increasing cranial capacity, a reduced face and teeth, larger body size. Did he create to mimic evolution and test our faith thereby?
>
> (Gould, 1984, pp. 122–123)

The evidence that Gould is giving here is from paleontology, where progression of transitional forms is evident to a "trained" eye. But if you present the fossils of half a dozen ape-like creatures to a group of school students and ask them to give their opinions, they might not give the same answer that Gould and other evolutionists would like to hear. Some of the answers that the students might give are as follows:

(1) One form evolved from the other.
(2) Each species is separate and there is no evolution, although there is apparent structural progression.
(3) They came from somewhere else, but we don't know where, so more investigation is needed.
(4) Somebody created them as structurally closely related.

Obviously, of all such answers, the evolutionists would like students to choose the first one. The reason is that:

> Biological phenomenon after biological phenomenon converges on evolution: Why are homologies so common? Why do we have such similar bone-structures between the functionally different arm of man, forelimb of horse, wing of bird, flipper of whale? Because of evolution! Why are embryos (e.g., man and dog) so similar, when adults are so different? Because of evolution! Why are the facts of biogeographical distribution so distinctive? Why would a group of islands like the Galapagos Archipelago have no less than fourteen different species of Darwin's finch? Because of evolution! These and many other phenomena put the actual fact or happening of evolution beyond "reasonable doubt."
>
> (Ruse, 1988, p. 118)

The readers can judge for themselves what the implication of the above paragraph would be if we replaced the phrase "Because of evolution!" by "Because of abrupt appearance!" or "Because of panspermia!" (the idea that life was originally seeded on Earth from space) or "Because of creation!" Is evolution more likely than panspermia, or panspermia more likely than creation? Who knows? But the great majority of school textbooks either promote the evolutionary view of life or omit a consideration of the question, and the job of a teacher is generally to teach the theory of evolution as scientifically true. The implication seems to be that as the matter has been settled—or is supposed to have been settled once and for all through the rigorous research of meticulous scientists to entertain debate on the issue in the classroom is a meaningless exercise. As a result, students learn the theory of evolution in the way that they learn Newton's laws of motion.

However, to force upon students one particular conclusion to the exclusion of others is to close their minds. What is as clear as apples falling to the ground is the existence of the "unbroken temporal sequence of progressively modern features." This is the fact. This is part of the world's data that Gould talks about. There is no argument among the scientists about the existence of this sequence. All credible and indeed non-credible scientists believe in this. The next question is how this sequence has come about. One explanation is that this sequence has come about through evolution. Evolution is the mechanism through which this might have happened. Can any scientist, or anybody for that matter, rule out the possibility that these species were not organically related? Certainly not. That these forms developed through evolution is only one possible answer out of many. It may be the most reasonable conclusion for many scientists, but it is certainly not the only conclusion. Evolution is not fact, but a way of explaining how that sequence happened, as are the theories of abrupt appearance and panspermia. All these truth claims have been subjected to scientific tests with different results, and the students should be made aware of this.

Nobody has witnessed evolution taking place. The process is far too slow for us to conduct a valid, scientifically verifiable experiment to prove or refute its reliability as a scientific fact. Thus, it is reasonable to doubt that this is the way new species arrive. However, evolutionists use the existence of skeletal structures that are closely related to older ones in the fossil record as a means of proving that evolution happened. But this does not bring us any closer to understanding the process of how evolution works. After all, if we cannot verify a certain claim that something has certainly happened, or at least if we cannot find an explanation for how this event happened, then it is reasonable to doubt the veracity of this claim. In 1835, it was thought that God had brought about the changes in the fossil record; later on, these changes were attributed to the developments of the older species in question. Again, the same problem arises. In the same way that we cannot find an explanation for the changes in the fossil record because no one has witnessed them, or provided a reasonable explanation for them, we cannot prove divine intervention through scientific investigations either.

Darwin too, felt the force of this objection. After the publication of *The Origin of Species*, this problem weighed heavily on his mind. The reality of evolution was being constantly impressed upon him by the multitude of facts that would be explained if it were true. But the trouble was that there was no proper explanation for his alleged process of evolution (Darwin, 1859).

Returning to the earlier discussion of the four answers that the students might have given, choosing the first answer would require an element of belief in evolution. To choose any of the above answers would require some kind of belief in what is known in science as a "hypothesis," which should be subjected to testing for its acceptance or rejection. As the first answer comes from a belief in evolution, the last comes from a belief in creation. A belief will have scientific status only if it withstands scientific tests. Contrary to what is presented in textbooks, evolution has not been conclusively proved as yet, even though evolutionists accept it as the only explanation of the forms of life that exist on Earth today. I am not ignoring the fact that for more than a century scientists have been genuinely researching in this field, but what I would like to emphasize is that they have not reached any conclusion as to where and how all living things have come from. There are a lot of disagreements among the evolutionary scientists regarding the exact status of the theory. This is further complicated by the research findings of those scientists who have explained the existence of all living creatures from a non-evolutionary perspective. Despite all these problems and uncertainties, evolution is taught with the zeal of believers, not much different from those who advance the theory of special creation. The parallels between the theory of evolution and creation have been noted by Colin Patterson of the British Museum of Natural History in an address given at the American Museum of Natural History on November 5, 1981:

> So in general I am trying to suggest two themes. The first is that evolutionism and creationism seem to have become very hard to distinguish, particularly lately. I have just been showing how Gillespie's bitterest characterization of creationism seems to be, as I think, applicable to evolutionism—a sign that the two are very similar.
> …
> So that is my first theme: that evolution and creation seem to be sharing remarkable parallels that are increasingly hard to tell apart.
> (Bird, 1989, p. 252)

This is where the teaching of evolution in schools has gone wrong. What are the flaws in the way we teach our children evolution? In a nutshell, despite the fact that there are a number of deficiencies in the theory, these deficiencies are rarely mentioned in the classroom, and the theory of evolution is all too often presented as absolutely true, with hardly any counterexamples or disagreeing data. Evolution has been presented as an indisputable fact and opposing views are ignored. We need to open up the minds of our students, so that they learn the truth as we have it now, and can question what they learn. This way they will learn to interpret scientific data rather than depend on purely conjectural imaginings. The criterion of "credibility" as a scientist should not be linked to one's belief in evolution or not.

The fact that there are deficiencies in the theory of evolution is incontrovertible; both evolutionists and antievolutionists agree on this—indeed, practically all scientific theories have deficiencies, although not to the same extent. The difference, however, is that evolutionists regard these deficiencies as difficulties that will be ironed out as our scientific knowledge increases, whereas their opponents maintain that the difficulties are sufficiently deep as to render the theory implausible. Given the indubitable existence of these inadequacies, they should be highlighted in the classroom, so that students know that there are flaws in the current theory of evolution. Textbooks should also present other competing theories supported by non-evolutionary scientists. I am not advocating the teaching of scientific creationism here, but arguing the need to present the views of those non-evolutionary scientists whose scientific integrity and scholarship in this field are beyond doubt.

Examples of how Evolution is Taught and Suggested Improvements
In order to show what I mean when I say that the deficiencies in the theory of evolution are rarely presented to school children, I will use the following textbooks as a representative sample of books written for age range 14–18 years and taught in British schools.

(a) M. Jones and G. Jones (2004a), *Biology—New Edition*, Cambridge University Press.
(b) B. Beckett and R. Gallagher (1996), *New Co-ordinated Science—Biology*, 2nd ed., Oxford University Press.
(c) G. Hill (1998), *Double Award Science for GCSE*, Hodder & Stoughton.
(d) M. Jones and G. Jones (2004b), *Advanced Biology*, Cambridge University Press.
(e) M. Kent (2000), *Advanced Biology*, Oxford University Press.

Jones and Jones (2004a) start their treatment of evolution by asking a question, "Where did all the different kinds of living organism come from?" (Jones & Jones, 2004a, p. 240) to which the answer given is:

> Today, the generally accepted idea is that the forms of life that now exist have gradually developed from much simpler ones. We think that life began on Earth about 4000 million years ago. Since then, more complex and varied organisms have developed—and are still developing. This is the process of evolution.
> (Jones & Jones, 2004a, p. 240)

This is generally the way evolution is introduced in school biology books. Another example is:

> Where have the millions of different living things come from? The most likely answer is that they were produced by *evolution* ... The first living things appeared about 3500 million years ago. They were no more than bubbles full of chemicals but they could reproduce. Some of the young were different from their parents. Some of the differences meant they could survive better than their parents. Over billions of years, these changes and improvements led to all the different creatures alive today.
> (Beckett & Gallagher, 1996, p. 54)

Such phrases as "the generally accepted idea," "we think that," and "the most likely answer" suggest that the idea of evolution is a contested one, but neither of the two books quoted above says so explicitly. Instead, these conjectural phrases, which vaguely allude to the fact that evolution is a controversial notion, are followed by definitive sentences that treat evolution as an accepted fact.

The students' confidence in evolution as an explanatory principle is then strengthened by the next section, 16.2, in Jones and Jones (2004a) that is rather boldly entitled *Fossils provide evidence for evolution* (p. 240). After presenting a sequence of fossil pictures from bacteria to mammals, the authors say, "One way of explaining this sequence is to suggest that more recent organisms such as mammals have developed from the earlier forms such as reptiles" (Jones & Jones, 2004a, p. 240). The student is then referred to a diagram of the tooth arrangements and differentiations of a sequence of animals, starting from a certain reptile (from 280 million years ago) to modern cats. The idea is to show a continuous progression from the reptilian arrangement of teeth all about the same size and shape to the mammalian one of teeth well differentiated into incisors, canines, and cheek teeth.

The sequence is clear enough, but to give a balanced view of the status of the fossil record, a number of factors need to be considered. For example, it is one thing to show how these animals can be laid out according to the arrangement of their teeth, but to extrapolate common descent or the transformation of species from this is a big jump indeed! A proper way to have presented this would have been to say that the given sequence might be taken as evidence for evolution, while helping students appreciate how difficult it is to draw conclusions about the overall biology of organisms from their skeletons alone, especially because the soft biology of extinct groups can never be known with any certainty. Teachers must highlight the pitfalls of drawing too many conclusions about extinct groups of which we have no information except skeletal.

Another example of this hazard is the issue of convergence. The skulls of the marsupial and placental dogs are almost identical, and without any other knowledge of these two organisms, one might conclude that they are closely related, yet in actual fact they are from two completely different groups and are very different in their soft anatomy, particularly their reproductive systems.

However, the idea that the fossil record provides the strongest evidence for evolution is echoed in Beckett and Gallagher:

> Fossils in younger rocks show an increase in complexity with time, and include an enormous variety of creatures which are different from today's plants and animals. Many fossils are of creatures which no longer exist. This is why the *fossil record* is the strongest evidence we have that evolution occurred.
> (Beckett & Gallagher, 1996, p. 56)

In Hill, too, we read:

> Scientists now believe that the first living things to inhabit the Earth were very simple organisms. Slowly, over millions of years, these simple creatures and plants have evolved into thousands of different organisms by natural selection ... The best evidence for evolution comes from *fossils* ... From the fossil record, a detailed picture can now be built up showing how animals and plants have changed since life began on earth and how they have adapted, evolved and sometimes become extinct over millions of years.
> (Hill, 1998, pp. 92–93)

Here again, one finds biology textbooks presenting evidence from fossils using such conclusive phrases as "the *fossil record* is the strongest evidence we have that evolution occurred," "simple creatures and plants have evolved into thousands of different organisms by natural selection," and "the best evidence for evolution comes from *fossils*." The presentation is so positive that we might expect there to be not the least shadow of doubt in the minds of students that evolution has occurred and the fossil record has proved it. Such statements ignore the controversies surrounding the sparse existence of intermediate forms and uncertainties in determining overall biology from skeletal structure alone that make it, some would argue, impossible to place meaningfully a particular creature in the sequence.

However, the way in which the paucity of the fossil record is mentioned in Jones and Jones (2004a) is interesting:

> Usually, there are big gaps in the fossil record, and we can only guess how the changes took place. The formation of a fossil is a rare event, and so it is not surprising that not enough fossils have yet been found for us to see exactly what happened to each kind of organism that lived in the past.
>
> (Jones & Jones, 2004a, p. 240)

This statement is made immediately after we are told that "we can find even more convincing evidence for this [evolution] by looking at a small section of the fossil record in detail" (Jones & Jones, 2004a, p. 240). Later on, in the next section, we find that "although the fossil record strongly suggests that evolution has happened, it does not actually prove it" (p. 244). It is to the authors' credit that in this respect they have provided a more balanced appraisal of the fossil record as "proof" of evolution than others.

However, the type of biased presentation as shown above is not unique to textbooks. Even in scholarly works one finds that the writings are biased in one way or another. For example, with regard to the paleontological evidence, Roger Cuffey wrote in 1984:

> Practicing paleontologists today, regardless of personal philosophical outlook, unanimously agree that the varied organisms inhabiting the earth originated by a process of gradual, continuous development or evolution over long periods of prehistoric time. Because the case for organic evolution had been adequately demonstrated in the late 1800's (principally by paleontologic evidence), scientists in this century turned their attention to many other important subjects.
>
> (Cuffey, 1984, p. 255)

> In summary, the paleontologic record displays numerous sequences of transitional fossils, oriented appropriately within the independently derivable geochronologic time framework, and morphologically and chronologically connecting earlier species with later species ... These sequences quite overwhelmingly support an evolutionary, rather than a *fiat*-creationist, view of the history of life.
>
> (Cuffey, 1984, p. 271)

As we will see below, such presentations are partial, and so do not give us the total story surrounding the issue of the fossil record. In order to have a balanced picture, it is necessary to consider the writings of other evolutionists who hold different views:

The evolutionist Gould says:

> The extreme rarity of transitional forms in the fossil record persists as the trade secret of paleontology. The evolutionary trees that adorn our textbooks have data only at the tips and nodes of their branches; the rest is inference, however reasonable, not the evidence of fossils.
>
> (Gould, 1977, p. 14)

Concerning the existence of apparently smooth sequences that do appear in the fossil record, such as the sequence between *Eohippus* and the modern horse, a favorite textbook example, Denton has the following astute point to make:

> If ten genera separate *Eohippus* from the modern horse then think of the uncountable myriads there must have been linking such diverse forms as land mammals and whales, or molluscs and arthropods. Yet all these myriads of life forms have vanished mysteriously, without leaving so much as a trace of their existence in the fossil record.
>
> (Denton, 1985, p. 186)

In terms of the way we present the theory of evolution in the classroom, why is this point not communicated to our pupils? Instead, we are faced with textbook after textbook that presents the sequence of fossils from *Eohippus* to the modern horse as "proof" of the way in which the fossil record supports the theory of evolution.

The evolution of human beings is also shrouded in mystery. According to Hill, an anthropologist at Harvard:

> Compared to other sciences, the mythic element is greatest in paleoanthropology. Hypotheses and stories of human evolution frequently arise unprompted by data and contain a large measure of general preconceptions, and the data which do exist are often insufficient to falsify or even substantiate them. Many interpretations are possible. These books all provide new alternatives, some refining the subject with new information; all, in varying degrees, supplant the old myths with new ones.
>
> (Hill, 1984, p. 189)

Surely, a more scientific approach for all textbooks would be to say that at the moment the fossil record does not really provide as conclusive evidence for the theory of evolution as evolutionists would like. In fact it suffers from a number of intractable problems that the evolutionists have not been able to solve as of yet. In order to provide a balanced view of the fossil evidence, authors could give a few diagrams to show some of the numerous cases in which there are enormous gaps between the first representative of a particular class and their nearest presumed ancestral types (e.g. Denton, 1985, pp. 167–171). Students must know, as Kitts has pointed out, that "fossils, by themselves, tell us nothing," and their meaning "must be reached within inferences that invoke biological theories" (Kitts, 1974, p. 458): the fossils can be interpreted in different ways by evolutionists and by non-evolutionary scientists.

We come now in Jones and Jones (2004a) to the section entitled *Evidence for evolution—homologous structures*:

> One way to explain the existence of these homologous bones is to suggest that all of these animals have evolved from an ancestral animal which had a "basic design" limb.
> The most likely ancestors for the amphibians, birds, reptiles and mammals are fish. One group of fish seem particularly likely candidates … It is quite easy to imagine that a fish like this could have been the ancestor of the amphibians, reptiles, birds and mammals.
>
> (Jones & Jones, 2004a, p. 244)

The speculative ideas expressed in the above quotation show how difficult it is to decide on the ancestry of animals on the basis of homology. In fact, there are deeper problems with homology, superficially persuasive though it is. Indeed, Darwin defines homology as the "relationship between parts that results from their development from corresponding embryonic parts" (Darwin, 1859, p. 492) However, in actual fact, embryology shows that homologous structures are sometimes arrived at by different routes. Denton, in his book, *Evolution: A Theory in Crisis,* describes a number of examples of the failure of homology (Denton, 1985, 142–156). He has shown, for example, that structures considered to be homologous in adult vertebrates cannot always be traced back to homologous cells or regions in the earliest stages of embryogenesis. The vertebrate alimentary canal is formed from the roof of the embryonic gut cavity in sharks, the floor in the lamprey, from the roof and floor in frogs, and from the lower layer of the embryonic disc in reptiles and birds. The vertebrate kidney is formed from the mesonephros in fish and amphibians, and from the metanephros in reptiles and mammals, two separate independently developing tissues. Even the vertebrate forelimb, considered strictly homologous, develops from different embryonic body segments in different organisms: segments 2, 3, 4, and 5 in the newt; 6, 7, 8, and 9 in the lizard; and 13, 14, 15, 16, 17, and 18 in humans, which shows that these are not strictly homologous at all. Other examples are the alimentary canals and excretory tubules of the larvae of insects, and the seeds of conifers and angiosperms.

The case for homology is severely weakened by the fact that homologous organs develop from non-homologous embryonic tissues. This does need to be addressed in schools, and the problem does not stop there. It should also be taught that the case is further weakened because homologous structures are specified by non-homologous genes; that is, different genes in the DNA of different species can specify homologous structures. The basic objection of some scientists is that homologous structures are not specified by homologous genes and do not follow homologous paths of embryological development (De Beer, 1971, pp. 15–16).

The situation is further complicated by the fact that almost every gene that has been studied in any of the higher organisms has more than one effect, a phenomenon referred to as pleiotropy. For example, in mice, the coat-color gene also has a secondary effect on the size of the mouse. In addition to this, the pleiotropic effects are species-specific. Almost every gene that has been studied in higher organisms has been found to affect more than one organ system. The problem that pleiotropy presents to homology can be demonstrated by discussing the multiple effects of one particular gene in some creatures (Denton, 1985).

The above examples illustrate some of the factors that undermine the validity of homology, but if we are to disregard homology completely, then how are we to explain why a fundamentally similar organ or structure (consider the vertebrate forelimb) is modified to serve often quite dissimilar ends? To my knowledge, no explanation other than descent from a common ancestor has ever been advanced for homology, and certainly creationists are at a loss for any reason other than that the Creator willed it so. For this reason, I believe that homology should definitely be taught in schools, but with all the caveats and qualifications that I have described above.

We now turn to the mechanism of evolution, that is, *natural selection*. In this regard, we come across the following in Jones and Jones (2004b): "But Darwin had no real scientific evidence or data to support his hypothesis of *natural* selection. Nor did he understand anything about the causes of variations, which are the essential basis on which natural selection can act, because at that time nothing was known about genes… Nor, despite the words in the title of his book, *On the Origin of Species*, did Darwin attempt to explain how a new species could arise" (Jones & Jones, 2004b, p. 147). This is a fair account of Darwin's position on natural selection. However, these authors go on to say that "Despite all of this, the essential ideas behind Darwin's theory of evolution by natural selection still hold firm today. Indeed, most biologists would say that evolution by natural selection is no longer a theory, but fact" (pp. 147-148). It is certainly true that most biologists accept evolution by natural selection as a fact, but what school textbooks do not mention is that there have always been renowned scientists from the days of Darwin who never accepted natural selection as being able to bring about the evolution of new species. In Jones and Jones (2004a), Section 16.5 describes the theory of natural selection as a mechanism that can explain how evolution happened, and Sections 16.9 and 16.10 give examples of evidence of natural selection in Kettlewell's peppered moths, and antibiotic resistance in bacteria. There is no disagreement among scientists that natural selection is operative in nature and plays an important role in *microevolution*, but scientists do not agree on its ability to produce new species:

There's no doubt at all that natural selection works – it's been repeatedly demonstrated by experiment. But the question of whether it produces new species is quite another matter. No one has ever produced a new species by means of natural selection, no one has ever got near it, and most of the current argument in neo-Darwinism is about this question (Leith, 1981, pp. 390-392). There are examples of speciation by natural selection as has been observed in European gulls, fruit flies (*Drosophila*), Hawaiian honeycreepers, wood warblers of North America, Caribbean island lizards, etc. However, these are all examples of microevolution where a gull remains a gull, a fly remains a fly, a lizard remains a lizard, and so on. What is disputed is natural selection's ability to explain the phenomenon of *macroevolution*, i.e., transpecific evolution. The degree of change that can be experimentally induced in organisms from bacteria to mammals has a limit beyond which no further change is possible. As regards the most popular textbook example of the peppered moth (*Biston betularia*), although Kettlewell himself considered it as "the most striking evolutionary change ever witnessed by man," Grassé, the most distinguished of French zoologists and editor of 28 volumes of *Traité de Zoologie*, said that peppered moths "either have nothing to do with evolution or are insignificant." The success of microevolution does not mean that macroevolution is true. Maze writes:

> In spite of the reverence that many systematists hold for neo-Darwinian evolution and natural selection, some biologists (Løvtrup, 1974; Rosen and Buth, 1980) doubt its merits, and philosophers have questioned the value of the theory of natural selection (Barker, 1969; Brady, 1979; Grene, 1974; Smart, 1963). Also, Macbeth (1974) has convincingly demonstrated the many inconsistencies in the arguments and writings of neo-Darwinists.
>
> (Maze, 1982, pp. 93)

What we are witnessing here is that there are many aspects of the theory of evolution about which there is no consensus among scientists. Despite such a lack of unanimity, students are only told that most biologists accept the theory of evolution, but are never explicitly informed that there are first-rate biologists who reject this theory on scientific grounds. For example, in *Advanced Biology*, a book for 16–18 year olds, "However, his [Darwin's] theory is supported by so much evidence that the majority of biologists accept it (Kent, 2000, p. 437).

Like most school biology books this book also does not mention, let alone discuss, the limitations of the theory of evolution. Interestingly, on page 436, in an inset entitled "creationism," it concisely describes that creationism is a belief in the Judeo–Christian doctrine that God created the universe in six days, and since their creation, species have never changed. It maintains that while many Christians believe that evolution is a way of God's creation, there are those who do not believe in the theory of evolution at all. This is a fair account of the stand of the Christians on this issue, and to dismiss creationism as an unscientific belief system may very well be justified, but in the same way, the true status of the theory of evolution should also have been presented with the same honesty and explicitness that it deserves.

The fields of paleontology, comparative anatomy, embryology, genetics, systematics, etc., do not provide compelling evidence for evolution. It was thought that the advances in biochemistry would solve the riddle of life, but Michael Behe, professor of biochemistry at Lehigh University, in his book, *Darwin's Black Box: The Biochemical Challenge to Evolution*, says:

> The scientific disciplines that were part of the evolutionary synthesis are all non-molecular. Yet for the Darwinian theory of evolution to be true, it has to account for the molecular structure of life. It is the purpose of this book to show that it does not."
> (Behe, 1996, p. 25)

School curricula should not completely overlook such controversies in the teaching of so important a subject as evolutionary biology. In the interests of educating our children with a fair and balanced view of evolution, they should be taught that the truth of microevolution is accepted among scientists but also that the success of microevolution does not necessarily mean that macroevolution is true. One should also incorporate into the curriculum the problem of organs of extreme perfection, which have puzzled scientists since the days of Darwin. Pupils should know that not all aspects of human behavior can be explained in terms of evolution. Evolution cannot, for example, explain our quest for knowledge, moral sense, and appreciation of beauty; many of our goals and ideals are not governed by survival urge or biological advantage. They should also be told that there are other non-evolutionary scientific theories on the appearance of various forms of life proposed by leading scientists from various fields of biology. I will not expound further on these, as I assume that readers will find it more useful to consult the original literature concerning them, but it may be noted that these problematic areas of the theory of evolution were not even mentioned in any of the five school textbooks quoted above, and this omission is by no means uncommon in other textbooks.

It should also be drawn to the attention of students that there have been opposing views to the theory of evolution from other disciplines as well. Students should learn that the discussion of the theory of evolution is not confined to the biological sciences, but because of what it touches upon, it generates interest among mathematicians, philosophers, sociologists, and lawyers as well. In this way, students will have an idea of how broad and wide-ranging the subject is.

Evolution and Mathematical Probability

As a rule, school textbooks do not discuss the mathematics of evolution, although this area has been studied in depth over the years. Mathematicians have looked at the problem from a purely theoretical point of view. Grave doubts were raised by a number of mathematicians and engineers at an international symposium held at the Wistar Institute that also included many evolutionary biologists. Schutzenberger of the University of Paris and MIT Professor Eden pointed out that the trial and error method—which is what is applied in the case of evolution by natural selection—is totally inadequate without the guidance of specific algorithms. At this symposium entitled "Mathematical Challenges to the Neo-Darwinian Interpretation of Evolution," Schutzenberger (1966) reported that the probability of evolution by mutation and natural selection is less than 10^{-1000}. According to Nobel-Prize-winning biochemist Monod (1972), the probability of life appearing on the earth was virtually zero. According to the molecular biologist Denton (1985), the probability that a single cell is self-assembled, which requires at least one hundred functional proteins to appear simultaneously in one place, can be estimated to be 10^{-2000}, which is infinitely small.

Stokes (1982) calculated that the probability of synthesizing a protein chain of 10^3 nucleotides is so small that the correct sequence would not be achieved in billions of years on billions of planets, each covered by a blanket of a concentrated watery solution of the necessary amino acids. An estimate of the probability of life originating by chance has been provided by the outstanding British astronomer Sir Fred Hoyle and the distinguished astrophysicist Chandra Wickramasinghe:

> By itself, this small probability could be faced, because one must contemplate not just a single shot at obtaining the enzyme, but a very large number of trials such as are supposed to have occurred in an organic soup early in the history of the Earth. The trouble is that there are about two thousand enzymes, and the chance of obtaining them all in a random trial is only one part in $(10^{20})^{2000} = 10^{40000}$, an outrageously small probability that could not be faced even if the whole universe consisted of organic soup.
>
> (Hoyle & Wickramasinghe, 1981, p. 24)

To these authors, "life had already evolved to a high information standard long before the Earth was born. We received life with the fundamental biochemical problems already solved" (p. xxii). The gist of such mathematical studies must be given to students so that they become aware of the standpoint of mathematics on this issue.

Conclusion

In this chapter, I have discussed the existence of the biased treatment of certain areas in evolutionary biology, citing examples from five school textbooks widely used in Britain. The space available here does not permit me even to list all the areas of debate that exist among scientists. Hundreds of books and articles have been written both in support of and against the theory of evolution. Just to give one example, in his most comprehensive work, *The Origin of Species Revisited*, W. R. Bird (1989) presents in depth both evolutionist and antievolutionist arguments from various fields of science. This book has no religious overtones, and it contains a balanced and meticulous presentation of the results of scientific investigation in the field of evolutionary biology and the theory of abrupt appearances with 2000 quotations and 5400 notes. It would be clear from a reading of such books that there are hundreds of non-creationist scientists from diverse fields of the biological sciences, many of whom are Nobel laureates, who have never been able to bring themselves to accept the validity of the theory of evolution. It is not their religious bigotry, but their genuine concern for scientific truth that led them to express their discontent with the theory of evolution: "In such a climate of scientific dispute, it is clear to all but the blinded zealots that Darwinism and macroevolution are not compellingly established or immune to criticism" (Vol I, p. 155). The attention of both students and teachers should be drawn to the existence of such materials and the continuing debate on the subject of evolution. We, as teachers, must resist the temptation of teaching the theory of evolution as a fact or as something that has been proved. We must accept the fact that mechanisms of macroevolution are not firmly established and as such do not preclude scientific alternatives:

> Macroevolution may legitimately be regarded as non-compelling if no plausible mechanism is offered to show that it can occur. Evolutionists are fond of saying that all agree on the "fact" of evolution, and are "merely" disagreeing on the mechanism. This is as odd as a trial, not on the assumed "fact" of the accused's guilt, but only on the "mechanism" of his assumed crime. Because macroevolution is not a "fact" ... and assertedly occurred by solely naturalistic processes, the mechanism remains vitally important.
> (Bird, 1989, Vol I, p. 155)

Our objective as educators should be to provide a balanced view of evolution to our children, by presenting to them a careful and fair evaluation of the points both for and against the theory. We should not ignore difficulties, anomalies, and alternatives, because they are needed if our children are to make a full assessment of the theory of evolution.

A measure of doubt entertained by a significant number of biologists in the theory of evolution should not be considered as some unfortunate irritant in the scientific process, but a necessary part of it. Science has its own limits, and a recognition of these limits is a must if science is to progress. Newtonian physics could not explain certain physical phenomena, as a result of which there developed new branches of physics, namely quantum mechanics and relativity. Resistance—although based on reasoned arguments—to an existing theory may appear outrageous in the beginning, but if we keep our mind open and are ready to delineate the limits within which our proposed theory comfortably works, then we will be able to avoid the kind of controversies that the theory of evolution has given rise to.

I have tried to show that there exists a debate in the field of evolutionary biology and that the debate is a genuine one. It has not been resolved during the course of one hundred and forty years, and there is no sign at the moment of its imminent resolution. Let us not pretend that the issue of evolution has been settled once and for all. Let our students not be brainwashed into believing only in the views of a particular group when in fact there is no complete consensus among the scientists. Instead, let us teach them the whole truth.

References

Beckett, B. & Gallagher, R. (1996). *New co-ordinated Science: Biology (2nd ed.)*. Oxford: Oxford University Press.

Behe, M. (1996). *Darwin's black box: The biochemical challenge to evolution*. New York: The Free Press.

Bird, W. R. (1989). *The origin of species revisited: The theories of evolution and of abrupt appearance, vols I & II*. New York, NY: Philosophical Library.

Cuffey, R. J. (1984). Paleontologic evidence and organic evolution. In A. Montagu (Ed.), *Science and creationism* (pp. 255-281). Oxford: Oxford University Press.

Darwin, C. (1859). *On the origin of species by means of natural selection, or the preservation of favoured races in the struggle for life*. London: John Murray.

De Beer, G. (1971). *Homology: An unsolved problem*. London: Oxford University Press.

Denton, M. (1985). *Evolution: A theory in crisis*. London: Burnett Books.

Gould, S. J. (1977). Evolution's erratic pace. *Natural History, May, 86*(5), 12-16.

Gould, S. J. (1984). Evolution as fact and theory. In A. Montagu (Ed.), *Science and creationism* (pp. 117-125). Oxford: Oxford University Press.

Hill, A. (1984). The scientists' bookshelf. *American Scientist, 72*(2), 188-189.

Hill, G. (1998). *Double award science for GCSE: Physics, chemistry, biology*, UK: Hodder & Stoughton.

Hoyle, F. & Wickramasinghe, C. (1981). *Evolution from space*. London: J. M. Dent and Sons.

Jones, M. & Jones, G. (2004a). *Biology: New edition*. Cambridge: Cambridge University Press.

Jones, M. & Jones, G. (2004b). *Advanced biology*. Cambridge: Cambridge University Press.

Kent, M. (2000). *Advanced biology*. Oxford: Oxford University Press.

Kitts, D. (1974). Paleontology and evolutionary theory. *Evolution, 28*, 458-472.

Leith, B. (1981). Are the reports of Darwin's death exaggerated? *The Listener, 106*, 390-392.

Maze, J. (1982). Neo-Darwinian evolution—panacea or popgun. *Systematic Zoology, 31*(1), 92–95.

Monod, J. (1972). *Chance and necessity*. London: Collins.

Raup, D. (1979). Conflicts between Darwin and paleontology. *Field Museum of Natural History Bulletin, January, 50*(1), 22-29.

Ruse, M. (1988). Is there a limit to our knowledge of evolution? In M. Ruse (Ed.), *But is it science? The philosophical question in the creation/evolution controversy* (pp. 116–126). Buffalo, NY: Prometheus Books.

Ruse, M. (1988). Witness testimony sheet: McLean vs. Arkansas. In M. Ruse (Ed.), *But is it science? The philosophical question in the creation/evolution controversy* (pp. 287–306). Buffalo, NY: Prometheus Books.

Schutzenberger, (1966). Algorithms and the neo-Darwinian theory of evolution. In P. S. Moorhead & M. M. Kaplan (Eds.), *Mathematical challenges to the neo-Darwinian interpretation of evolution. A Symposium held at the Wistar Institute of Anatomy and Biology, April 25 and 26, 1966*. Philadelphia: The Wistar Institute Press Symposium Monograph No. 5, published by The Wistar Institute Press.

Stokes, W. (1982). *Essentials of Earth history: An introduction to historical geology (4th ed.)*. Englewood Cliffs, New Jersey: Prentice Hall, Inc.

Fundamentalist and Scientific Discourse: Beyond the War Metaphors and Rhetoric

Wolff-Michael Roth

At the time that Jesus Christ was born, there was a population of a million—I am just trying to think, so it must have taken a lot of time for the population to get really big, so [the Earth was] probably [created] about 50,000–60,000 BC, that's what I believe. I am not sure what the dinosaur bones mean and all that. I just don't find that logical how God would actually spend the time and make the actual forms on the Earth and stuff like that. I find, he said, "Let there be land." He didn't think of every little detail, the Grand Canyon or the Rocky Mountains. It just does not seem logical to me that God would actually do that. Whereas science does seem logical in that aspect, having the concept of erosion and all that.

(Preston, 17-year-old high school student)

Saying that what we learn shapes our understanding has become commonplace; it has equally become commonplace to state that our understanding (scientific or naïve theory) shapes how we perceive the world. Because students are not blank slates, their existing practical (tacit) understandings and intuitions of how the world works will mediate any learning and development they experience at school. Having emerged during prior experiences, these understandings and intuitions constitute a background against which new, school-related experiences become salient.

In the past—when students talked (about) science, such as forces or atoms—any deviations from scientific discourse has been considered as misconceptions, alternative frameworks, or naive conceptions. When students' talk pertained to something apparently related to how the mind works, epistemology, God, or creation, it has been discussed in terms of some belief that is thought to interfere with learning the scientific canon. Thus, science educators may read the introductory quote and say that Preston held certain beliefs, about God, creation, and the age of the Earth; these beliefs are thought to reside somewhere in the mind of the person. They might point to studies suggesting that such beliefs correlate with levels of abstract thinking and interfere with being able to accept the theory of evolution (e.g., Lawson & Worsnop, 1992). However, a sound theory of knowing that explicates the relationship between religious beliefs and scientific knowledge has, to my knowledge, not yet been provided. It would, anyhow, have to deal with the fact that students have to take not only religion but also science on faith—students in my studies on epistemology and the nature of science have remarked that their textbook does not differ from the bible. Both forms of text have to be taken on faith.

In the quote, we also see that Preston did not only make statements but also enacted rough calculations and evaluated different claims. First, he estimated the Earth's population at the time of Christ to have been about a million, and inferred that creation occurred about 50,000–60,000 years rather than the previously talked about fundamentalist Christian claim of 5000–6000 years. Second, he described the concept of erosion as a more plausible source as a shaping force of the details of the natural world than the claim that God had thought of and created all these details. That is, the quote constitutes evidence that Preston does not just talk about some accepted versions of the world but exhibits reasoning processes that are normally associated with science. It is not, therefore, that (highly) religious students close themselves to rational considerations; many strongly religious students do reason in ways characteristic of science (Roth & Alexander, 1997). The real issues lie somewhere else.

In fact, we might ask, what are the processes by means of which such statements about God and creation are supposed to interfere with the learning of scientific facts, concepts, and theories? Some scholars have suggested that students might have knowledge about something (evolution) without believing in it (Chinn & Brewer, 1998). The issue of how a belief can interfere with the acquisition, appropriation, or construction of new knowledge is therefore still to be settled. Whereas students' everyday (nonscientific) descriptions of physical events have been well studied over the past quarter of a century, other experiential sources and discourses including ethical, religious, or emotional issues are less charted. Thus, although scientific and the religious discourses of fundamentalist Christians have often been incommensurable at the institutional level (e.g., the creation vs. evolution debate in the US), religious discourse has rarely been studied as a mediating element during science instruction and learning processes.

We can also take a very different perspective on the opening quote by assuming that rather than exemplifying Preston's beliefs and reasoning, the text is the contingent outcome of an interaction between him and the interviewer, and therefore bears all the marks of a collective rather than individual achievement. Furthermore, because Preston and the interviewer use language, what is being said does not just characterize the interaction of the two but in fact exemplifies a concrete possibility of talking about the age of the universe, God, evolution, or erosion that exists at the collective level and therefore reflects as much cultural–historical as it reflects individual ways of talking. What has been said can therefore not be attributed to Preston alone, or to Preston responding to the interviewer and his questions, but has to be understood as part of a dialectic that bears the mark of individual and collective. We do not need to posit beliefs and knowledge to understand what Preston has said but can simply assume that people constitute relevant beliefs, knowledge, and reasoning in relevant situations drawing on words, gestures, body positions, artifacts, and other salient resources.

Therefore, the quote can be analyzed in a different way, based on which resources have been rallied in the constitution of knowledge and beliefs. Thus, the probable age of the universe was established by an inference from an estimated population and the amount of time it might have taken to develop from two ancestors (Adam and Eve). Preston drew on the repertoire of rationality (Roth & Lucas, 1997). Invoking God is an instance of deploying a religious repertoire; by rejecting the point that God created the specific shapes of Grand Canyon and Rocky Mountains, Preston employed the repertoire of intuitive truth. The same intuitive repertoire is deployed to support the usefulness of science ("Science does seem logical ... having the concept of erosion"). The advantage of taking a discursive approach is that it allows us to understand the varied ways in which knowledge, beliefs, truth, or claims are constituted in situation and as a function of the ongoing activity. Different situations, conversational foci, or activities are associated with different salient resources, making it more likely that knowledge, beliefs, facts, and truth claims are different rather than the same from situation to situation. Researchers, therefore, have to provide empirical evidence supporting the existence of stable knowledge, conceptions, beliefs, etc.

In this chapter, I focus on why Preston (or a person like him) would articulate issues involving science and religion in the way he did rather than on the nature of the belief he articulated. Because every (discursive) action not only produces something (an object, a sentence) but also produces the speaking subject, I am also concerned with the question of identity and the narrative elements used to articulate self and (social) other in lived experience and self and world as mediated by knowledge. The former relation involves an experience of bodily presence, whereas the latter has only a past and no presence. Here is where we can articulate the fundamental difference between science and religion: they constitute different modalities of time and being. Science is about objects, which consist in having been; religion is about what is, experience, and grace, which consist of being in the present to which we may add a future orientation, as apparent in revelation and being toward death.

Thinking about the difference between science and religion, I am fundamentally concerned with the nature of language and how it is used to manage the relationship of the two. This leads to three interrelated issues to be addressed here. First, language is the central tool for constituting the existential issues and, therefore, who we are; however, although (auto-) biographical accounts seem to be about the singularity of an individual, they are always also about a character and character traits, which are collective (shared), general images of what humans are like. Any autobiographical discourse is, therefore, a concrete realization of collective possibilities of autobiographical discourse. Second, because language is a central aspect of who we are (rather than a mere representational tool), the conflict some students (people) feel between religion and science goes beyond mere cognitive issues (knowledge, belief); the real issues are existential, reaching deep into the fundamental question of who we are, individually and collectively. Third, in a reflexive manner, language is also the tool for the interaction between an interviewer and the research participant; the interview product (recording, transcript), therefore, bears all the marks of the interaction and cannot be attributed to the interviewee alone.

Because we use language, our (auto-) biographical narratives are always dialectical, both about ourselves and not about ourselves, about who I am (particular) and about how one can be (general). Any (auto-) biographical narrative, written in some medium or emerging from an interview, therefore, develops characters (one that the author may identify with) whose actions are intelligible because they reflect a type rather than a singularity.

Language, World, and Identity
Dialectical Nature of Language
A growing body of research suggests that substantial theoretical and methodological difficulties emerge when we treat language as a pathway to attitudes, beliefs, conceptions, ideas, knowledge, or memory (e.g., Lynch & Bogen, 1996). Another body of research has articulated the theoretical difficulties that emerge when language is treated as a representation of (some aspect of) the world (Latour, 1999). A different, pragmatic approach to language regards it as a means for doing things, as a medium of communication, tool in social transactions, way of maintaining and further weaving and reweaving of the network of human relations that makes society (Rorty, 1989). In such an approach, there is, therefore, no more difference in knowing a language and knowing the world more generally—language inhabits us and is our habitat. Attitudes, beliefs, conceptions, ideas, knowledge, or memory are merely reified outcomes of situated interactions rather than distinct objects or qualities in the minds of people.

Language has a dialectical nature, which French philosophers appreciate in the relationship of *langue* and *parole*. *Langue* denotes language as a generalized (abstract) system, universally accepted within a community; *parole* denotes the actual linguistic behavior of people. The relation of *langue* and *parole* is dialectical because every instance of language use (*parole*) constitutes a concrete, particular realization of possibilities that exist at the generalized, universal level (*langue*). The language I use to write these lines, this text, is therefore both mine and not mine—which led Derrida (1998, p. 2) to write, "Yes, I only have one language, yet it is not mine."

Consciousness, including self-consciousness, requires social mediation—which implies actions and tools to reflexively make actions the topic of further actions (Mikhailov, 1980). Language is our main tool that achieves both of these functions; but language alone is insufficient to explain the different ways in which we account for our experience. If human beings were the sum of their nameable traits, they would be no more than machines, subject to the program inscribed in traits, memory, etc. But human beings are active molders of their situation, coparticipating in social interaction and thereby actively producing and reproducing themselves and society. Agency is embodied in the character, the human counterpart of the narrative plot of novels and plays. We understand ourselves in terms of characters in plots. When we asked for beliefs, wishes, desires, goals, and so forth, we provide narratives in which we appear as characters making rational decisions. Thus, "Life must be gathered together if it is to be placed within the intention of genuine life. If my life cannot be grasped as a singular totality, I could never hope it to be successful, complete" (Ricœur, 1992, p. 160). This gathering is done in and with (life history, autobiographical) narratives. In these narratives, actions are sensibly and reasonably related to activities; in fact, giving an account of some action can be understood as a way of describing the rationality of actions.

Language, Narrative, and Self

Who we are can never be independent of language. My language "constitutes me, it dictates even the ipseity [personal identity, individuality, selfhood–WMR] of all things to me, and also prescribes a monastic solitude for me; as if, even before learning to speak, I had been bound by some vows" (Derrida, 1998, p. 1–2). Who we make Preston out to be is a result of his interactions with the social and material world, here with an interviewer talking to him about science and religion; Preston comes to be articulated in terms of entities that surround him and the interviewer, intelligible to both—Jesus Christ, God, Grand Canyon, science—and that one or the other may reject as existing. This plays into the hands of a pragmatist conception, which accepts that there are no private languages (Wittgenstein, 1968)—language is always of the other, from the other, for the other. Who we are when we talk about ourselves and our world (beliefs, knowledge) is therefore always and already a social way of talking, we are always and already who one can be in a given culture, "For the I of the basic word I-You is different from that of the basic word I-It" (Buber, 1970, p. 53). We are therefore both living beings (marked by the I that confronts You) and characters in plays (the I surrounded by a multitude of contents), in narratives that others can understand. In the former situation, "love occurs," in the latter, "one 'has' feelings"; in the former situation, "feelings dwell in man," in the latter, "man dwells in his love" (p. 66). That is, Preston appears twice: he is both the person confronting the interviewer (with etymological origins in "seeing each other" and "having a glimpse of"), who refers to himself as "I" in an "I-You" relation and the person referring to himself in the past, in an objectified form, in an "I-It" relation to the surrounding things.

The accounts human beings provide of anything are functions of the situation—coparticipants, resources (language, salient environment, motives of activity). It therefore makes little sense to speak of Preston independent of the situation; the Preston who emerges from the transcript is the result of the transaction with the interviewer, the language that was salient, the topics of talk, etc. "Self-constancy is for each person that manner of conducting himself or herself so that others can *count on* that person. Because someone is counting on me, I am *accountable for* my actions before another" (Ricœur, 1992, p. 165).

In the narrative that unfolds, Preston appears not in his singularity as individual, he is not just the "I" in the "I-You," but as a concrete realization of the possibilities to be a person in his society at that time ("I-It"), who comes to be understandable exactly because of a character and character traits that he shares with others in the culture, which also provides the linguistic resources that name and therefore bring to life these traits.

Science and Religion as Modes of Being

As a teacher and scholar, I am concerned about a different way of dealing with interview texts and therefore with the issue of being in the world, language, and the relationship of science and religion. Here, I exemplify my approach, which leads to rather different classroom implications then one might get from other approaches. In the following, I provide an exemplary analysis of interview material. My analyses make salient (a) the nature of interview texts as outcomes of collective activities; (b) how language allows an individual to take up a place in the world; (c) the double dialectic of character/plot and individual realization of cultural possibilities; (d) the difference between cognitive conflict and experience of conflict; and (e) an autobiographical text as a concrete realization of cultural learning, development, and change.

I draw on interview material that derives from a longitudinal study of students' talk about epistemology, nature of science, and learning. Over an eighteen-month period, the students who were enrolled in two consecutive physics courses read and discussed in class a variety of texts on epistemological issues. They wrote essays, participated in interviews regarding epistemology, the nature of science, and learning. Class discussions, interviews, and essays were transcribed, leading to a database encompassing about 3500 pages of transcripts. As a teacher–researcher, I taught the physics courses, collected the data, and prepared all transcriptions. In the analyses that follow, I use a third-person discourse to refer to myself, because it allows me to attain a distance (Buber's "I" surrounded by "contents") necessary to enact a critical analysis and deepening of understanding. In particular, it allows me to more easily separate my public actions—available to others, Preston and readers of the transcript—from my intentions—neither accessible to others (then and now) nor, in their original form, to me today.

In his last two years at school, Preston had been a moderately successful student; his grade point average was about one-half standard deviation below the mean. He was less successful in his two sciences, chemistry and physics—he narrowly passed the first and failed the second. Preston had enrolled in the sciences in part because he understood his parents as wanting him to become a medical doctor. Senior-level physics and chemistry were university entrance requirements for entering the pre-med science programs. Preston had a keen interest in theater arts (he received an A grade) and was actively engaged in several school plays. During the two years at the school, he repeatedly talked about his deep religious commitments and the conflicts he experienced as he learned chemistry and physics. His peers also knew him as an avid debater with respect to religious issues. Preston was a chapel warden and member of the chapel choir.

Interview Texts as Outcomes of Situated, Collective Activities
Interviews are generally taken as direct pathways to knowledge, beliefs, motivations, and emotions, all of which are thought to reside in the heads of students were they might come into conflict, enhance or interfere with one another, and thereby change how students learn. However, any interview presupposes a set up—an interviewer who asks questions and an interviewee who responds—and the use of language—an interview without some form of language is an oxymoron, in the modern use of the word. It is a situation to which participants orient, where they find themselves and language, and which constitutes a way of being, where "I" is bodily confronted with "You." The details of the situation, as understood by the participants, the language used, and the transactions between interviewer and interviewee inherently constitute interactional resources that structure the outcomes—the interview text as recorded and transcribed and "I" becomes an "It" and is "surrounded by a multitude of 'contents'" (Buber, 1970, p. 63). In the following episode, the interviewer frames the topic of the conversation, the items of a survey to which students had responded by taking a position (agree, disagree, other) with respect to each of five statements (agree, disagree, other) and then explaining their choice. The first statement to be discussed is, "Scientific knowledge is artificial and does not show nature as it really is."

Interviewer: I want to know from you what your ideas are about these five questions from the survey and if you change your view just let me know what sort of caused that change. Because I am going to a conference about the history and philosophy of science in science education and there are people who argue that what one should talk to students—you know, students should get an appreciation of these sorts of discussions. Why don't you tell me what you think about that first statement, "Scientific knowledge is artificial"?

Preston: Well, first of all, I am a very religious person and that sort of rearranges my knowledge about science, how I think about different things. When I think back of all the things I have been taught such as how the Earth was made, how things are developing, it makes me think about what is true and what isn't, about what is real and what isn't, and sometimes the fact that what I have heard in science and religion can be compared to each other.

In his framing, the interviewer accounts for the fact that he has invited the student for an interview by stating a broader societal motive, a "conference about the history and philosophy in science education" where there are "people who argue that one should talk to students ... about these sorts of discussions." In this way, the interviewer sets up the particular talk (action) in relation to a broader motive (interests of a scientific community).

Preston is then asked to talk about what he thinks about a particular statement, "Scientific knowledge is artificial." However, the student does not answer the question directly but rather produces a contextualizing frame, in turn. Although the original question was about science, the response has to be understood in the context of a narrative of self who is also a religious person. The answer is also framed in terms of what he has heard in the past—memory work is being done here—and that those things he has heard are sometimes addressed by both science and religion. Although the interview protocol had not foreseen questions about the relationship between science and religion, the topic, once it had emerged, is unfolded and carried by both participants.

Preston introduced the contrast that features him as a source ("it makes me think") and recipient of agency ("the things I have been taught"). In fact, the phrase "it makes me think" already contains the dialectical tension between a person who is both agent and recipient of agency, subject of and contributor to life conditions. The character developed is also subject to different messages ("all the things") that have to be evaluated in terms of their truth-value ("it makes me think about what is true and what isn't").

"Being religious" is not something that is singular to Preston: In saying "I am a religious person," he is providing a description both of himself and of a way of being in this culture more generally. That is, "being a religious person" is both a generalized possibility *and* a concrete way of being. That is, "being religious" is the description of a particular character trait that is ascribed to different characters in (auto-) biographical narratives. The very statement "I am a very religious person" is marked—for it to make sense it implies a non-articulated horizon of others who are not so religious or are completely without religion. This, then, is one of the contrasts that have to be managed by the character developed in the narrative, carried by both interviewee and interviewer. Thus, the interviewer will be seen to return to the contrast, requiring the responsive interviewee to articulate issues even if he had never pondered or discussed them.

A View of Discourse as Part of Being in the World
Language is a central aspect of being in the world rather than an independent, value-free medium for representing and externalizing thoughts, conceptions, knowledge, and beliefs. This becomes evident in the following episode where Preston continued by providing the example of different versions of the origin of the world. He introduced a disjunction that associated science and atheism and contrasted it with the beliefs of the religious person, here concretely

realized in his own person. Further, the greater extent to which science explains the world is linked to the ease with which one can believe in it. Preston set up a contrast between competing versions of the origin of the Earth. The scientific version (e.g., "dust collected out of space") appeared more fruitful than religion (e.g., "explains the world more than religion"), perhaps because of the continued process that it articulates rather than spontaneous creation. This contrast, in fact, is set up in support of the claim that science is not artificial—the question that was the first topic in the interview. The account then suggested that more people are drawn to science because it is easier to believe than religion, its presuppositions easier to accept. Here, both religion and science appeared as sets of beliefs, which make competing claims about nature and the universe. Preston then made an existential claim, about people who are attracted to science—atheists. An atheist mind is more easily attracted to science than a religious one. In the end, he sets up the dialectic of individual as agent ("I sort of understand") and sufferer ("I am drawn").

The issue, then, is not merely one of beliefs or cognition, but it is one that involves a person as a whole. "I am very religious" is a statement about a way of being in the world rather than about something that can be exclusively relegated to the (cognitive) mind. The issue is further elaborated in the next episode, where Preston described himself as a person who is "caught in the middle." The question of science and religion and the conflicting versions that they seem to provide about the origin is something that affects the person.

Preston: Well I am sort of caught in the middle, because I have grown up with religion. But I have also grown up with science. I guess I really can't determine what is right and what is wrong. If I were an atheist and not believing in God at all, I would automatically go towards science. I would say, "Science is this, science is that," and then people say, "You know, science can actually provide the information about what is and what is not."

The interviewer had asked Preston to say something about the contrasting claims that (fundamentalist) Christians and scientists make about the origin of the universe—(spontaneous) creation versus big bang theory. In response, Preston articulated himself as a character caught in the middle between different versions of the world provided by the two systems, science and religion. This character finds it difficult to determine "what is right and what is wrong," which of the two versions to accept. He then develops a different plot, one in which the character is an atheist, who, because it implies a situation without conflict, "would automatically go towards science." For this person, there would be no conflict, there would only be science; and others in this plot would confirm, asserting that science actually provided the information about what is and what is not, an answer that is not available to him because the conflicting statements are made by different authorities (science, religion).

In this plot, there are no differences within a system (science, religion) but only between them. The character in the plot may not be able to decide. Preston provided different versions of himself: he develops a different concrete character, which is also a realization of such a character generally.

The interviewer then asked Preston how he experienced this contradiction—whether he tried not to think about it or whether these differences are only salient sometimes ("on and off"). When the student states that this experience is present at all times, the interviewer pursues whether he was worried by it. (Elsewhere in the interviews Preston stated that the experience is like a hit to the face every time he enters a science classroom.) In response, Preston further articulated the issues in existential terms.

Preston: It distorts me. It really troubles me a lot. Because if I think about science, I feel like I am drawn away from religion. That really worries me a lot, because I feel like I am being taken away from what I have been a part of, which is religion. If I go toward religion, I feel like I am not giving science a chance at all; and I can't see myself doing that because I am a person of morals.

It is not that he merely has to rationally evaluate competing claims, but he, the person, is drawn away from religion, something that he has been part of, and which is a part of him. But if he "go[es] toward religion," then he feels like he is "not giving science a chance at all."

Going this way or that way involves the whole person, the one who, as the protagonist in Robert Frost's famous poem "The Road not Taken" ponders, always has to choose one road over another, never knowing what would have been had he taken the other. The person in Preston's narrative cannot choose science over religion or religion over science, because the moral person has to give a chance to all sides of an issue. That is, the conflict between science and religion is not just something that could be resolved by abandoning one for the other, because the moral person has to ponder all versions equally ("give a chance to"). Here again, the "moral person" is a generalized possibility to be in the culture, and Preston's discourse reifies the person referred to with the reflexive "I" as a concrete realization of this character.

There would be no room for a tragedy of having to choose unless the universalist approach of the moral person giving a chance to science and religion and the contextualist approach to the absence of conflict within each system had to be maintained in place, and unless the practical mediation capable of surmounting the antinomy were entrusted to the practical wisdom of the moral person's judgment. The moral person finds himself at the crossroad of competing discourse, competing ways of being in the world, and has to make a choice—even not acting (choosing) is an act, which, here, prolongs the effects of the antinomy.

Dialectic of the General (Character, Plot) and the Particular (Individual)
Until now, the conflict between science and religion, in general, was articulated in terms of a conflict between scientific and religious discourse regarding concrete issues. As part of the previous exposition, Preston has evoked two types of people, the atheist and the religious person associated with the conflicting fields that constitute them. (This involves the same dialectic as character and plot.) In the following, Preston couched the issues in narratives of debates with peers, narratives in which the religious person is facing atheists in antagonistic fashion. The conflict between science and religion becomes rearticulated and recontextualized as a conflict between adherents of different belief systems.

Preston: I talked to people who are complete atheists, who think that they are really good [at] science, that they want to be scientists. I said, "What about religion, have you ever thought about what you actually want to believe, have you ever thought about what your destiny is? Science can't provide you that." All they can say is what you are made of. They can't say what you are going to be, where you are going to go. Haven't *you* [interviewer] ever come across any Christians at all? Any person who, as I do, believes in both religion and science?

This unfolding narrative is about a debate, involving the religious person ("I") and complete atheists pitted against one another in argument. (Elsewhere in the transcripts, Preston states that the interviewer is likely an atheist because he is also a science teacher.) The religious person asks whether his opponents have thought about issues that involve the person, his beliefs and his destiny, what his future will be, and what the end point of life's trajectory will be ("where you are going to go"). All of these are questions that involve the person. Contrasted with this is the response from the atheists, who state matters of fact ("what you are made of").

Until now, the atheist nature of others in this school has been an unsupported statement. Preston later provided supporting evidence. During the daily compulsory ecumenical morning service, he had noticed others ("every person") in a demeanor that is not normally associated with the devout person in respect, standing before his God ("head up like this"). "They look around," implies interest in and attention to something other than reflection and prayer. Because they are atheists, these others do not experience conflict between different discourses; they do not find themselves at a crossroads where they have to make the difficult decision between science and religion. This implies that science is their future; in fact, they need a future science in the same way that a religious person needs a future in his faith. Because these students are atheists, they need science, perhaps to fill what religion is to him.

In his next turn, the interviewer raised the issue that there are individuals, scientists, who "have resolved these issues for themselves." That is, the interviewer suggests characters and plots where a person can be a scientist and a religious person simultaneously, persons "who accept big bang and believe in evolution."

These images and forms of reasoning are not singular to Preston; these are forms of being, thinking, and describing that exist at the collective level. At the same time, the conflict involves the whole person, something he is wrestling with rather than being something that could be decided logically, by a disembodied mind, independent of who we are. It involves the person in a concrete and practical way; it pervades his life ("something I have wrestled with for a long time"). In fact, the interviewer talked about the issues in terms of people who come to grips with different versions of the world, who live both worlds at the same time; he did not set knowledge (claim) against knowledge (claim) that could be evaluated by a computer program.

Cognitive Conflict and Experience of Conflict
Cognition is often presented in terms of the processing and transformation of information independent of the particular entity (computer, mind) that implements the cognitive processes. Emotions and motivations then have to be introduced as external processes that somehow interfere with thought, which thereby is raised from cold to hot cognition. In the approach taken here, narratives involving our selves are never mere reflection of thoughts, but constitute ways of being in the world; these ways are about people as a whole rather than about disembodied knowledge and thoughts.

The interviewer had made a reference to scientists who have resolved the potential conflict between alternative versions of the origin of the Earth by suggesting that such a resolution was something for the interviewee to wrestle with. In response, Preston suggested that this is an issue that he had "wrestled with for a long time." The interview topic then took a new turn as the interviewer asked the student (who apparently expected the question to pertain to his "think[ing] *about* science") about the effect this (wrestling) has on what he can do in science.

Preston: Why do you think it is hard for me to grasp the concepts in class, hearing you lecture in class about the way light comes in and how it diffracts into an electron? It's hard for me to grasp that knowledge, whereas complete atheists, they grab it and say, "I can agree with you [interviewer]." I was thinking, "How can you actually say that when you know other things, like what religion has to offer?" Because I am not a very full Christian, I cannot say, "Well, God has created life, that's the way life is." I can't say that because I don't know much about what the bible is to other people. But there have been times when it has affected me, it's sort of reflected in my marks. I never really told anyone, because they wouldn't actually believe me if I was to say, "Well, I can't take chemistry, because I don't believe in that. I just do the chemistry to get the marks." But I feel guilty on the other hand because that's not what I believe in.

Here, Preston initially developed the trait of the religious person, who finds science concepts hard to grasp, and contrasts this with the atheist, who can grab hold of them, and therefore finds it easy to agree with the teacher. He elaborates, how a person could actually accept and agree with something "when you know other things" such as those things that "religion has to offer." He then articulated himself as not being a sufficient expert or authority ("not a very full Christian") who could contradict the teacher or other students. He described that this (continuing) conflict is expressed in his marks and that even explaining the conflict to others would not help "because they wouldn't actually believe" him. Despite not believing in its content, he took chemistry to get the number of senior-level courses ("marks") required for university entry. Taking chemistry, although he did not believe in it as a body of knowledge, led him to feel guilty.

This is the narrative about a religious person coming face to face with a discipline characterized by a body of knowledge at least some of which contradicts understanding of what religion has to offer. The main character finds himself in a situation where he cannot contradict the statement of others because he feels insufficiently qualified; and he cannot tell others about the conflict he experiences because these others would not believe him. The plot exposes the character to an almost unbearable conflict, which he can neither resolve nor talk about. The character takes on the burden of the situation, with the result that he feels guilty by doing something that he does not believe in.

In asking whether his "parents, the church, or whoever gave [him] a lot of guidance," the interviewer appears wanting to explore the origin of the religious commitment. This question turns out to have been an invitation to the elaboration of a narrative about how Preston became the Christian that he is. Initially, religion was a mediating element in the conversion of a brat to a calm, reasonable, and altruistic person, "someone who listens and really understands how other people feel."

The narrative articulates more conflicts, a shy protagonist who did not talk very much and who did not want to talk physics as language (a reference to *Inventing Reality: Physics as Language*, a book students read in their physics course). Conflicts are evoked, which had first arisen when Preston had come home in sixth grade from a science class and debated with his parents on scientific versions of events that conflicted with their religious views. As a consequence of their guidance, he "started to lean more and more toward religion." But now in school, far from his parents, he attends physics and chemistry, which makes him "start to believe in that more and more."

This is the narrative about an individual who does not give into peer pressure, who follows his own beliefs. But it is also the narrative about someone being influenced by teachers and pastors, authorities in the respective fields.

The interviewer continues to pursue the question of conflicting evidence coming from the two domains, science and religion. Here, the conflict was framed in the form of making a decision about the truth-value of different texts.

Preston articulated a plot of a person whose early education biased him in one rather than the other way. He articulates believing in the truth of the bible, a consequence of growing up as a religious person. In the alternative plot, growing up as an atheist or scientist, he would believe in the science textbook. His parents were the influence that made him go one way rather than the other. In this narrative, the parents actually provided a reason for influencing him—religion and God are dependable resources for situations when you are in trouble (as when he was a "brat") or despair. Here again, the issue pertains to the whole person in terms of a character in a plot—trouble is not something at the cognitive level but a situation for the person. Preston describes himself as a "brat," which got him into trouble at school. The trouble is real and experienced, not just a mental state.

Autobiography: Learning, Development, and Change
The autobiographical nature of character and plot that are developed in such interviews allow the articulation of learning, development, and change narratives. Again, rather than being narratives about particular learning, development, and change, these narratives are concrete realizations of possible narratives that exist for and can be drawn upon by all members of a culture speaking the language.

The interview text continues to plot the character as being subject to external influences. As a child, parents and the church constituted the sources of these influences. Now, in high school, he is exposed to other sources (e.g., science) that not only influence him but also create the conflict. When he asked for assistance, for a miracle like those that he read about in the bible, God did not give him a sign. Yet, he continued to follow the path of religion because of his parents (early) influence on who he is, a religious person.

Interviewer: What makes you believe that what is in the bible is true and what you see in a science textbook is not true? What is the difference between the bible, which is a book, and a textbook, which is also a book?
Preston: The only explanation that I can give for that is that I have grown up as a religious boy. I am sure that if I had grown up as an atheist, I would probably question all this. If I had grown up as a scientist, I would question all that the bible is about as well. I just had grown up in a way, and my mind was focused like that. My parents have forced me to believe that God was the only thing above Earth that you can count on, that you can depend on if you have trouble or when you are in despair.

Preston introduces a trait that allows a character to change over in his beliefs—being not systematic. Such a character can "go completely against" religion and pursue the path of science. Preston describes a character type, for it allows Christians to convert and become Buddhists and allows atheists and heathens to convert and become Christians (e.g., St. Paul). Being torn by the different forces of science and religion, therefore, is the expression of a more general character trait, perhaps pervasive to all human beings, of being not systematic, incontrovertible to the bitter end, and therefore open to conversion experiences.

Accounting for something is a way of articulating a background against which what someone says becomes reasonable. Thus, we do not know whether Preston has ever pondered the question about the relative truth-value of different kinds of text, the Bible and science textbooks. We have to understand the response as a situated contribution to the unfolding topic with the resources at hand. In this case, we are not able to postulate a model in his mind that he represents in language and thereby makes available to the interviewer and others.

The following, final episode reveals evidence for change that is possible. The interviewer has once again raised the question of truth, now relating to the question of the existence of God. On the one hand, Preston seems to question the existence of God and, on the other, he underscores his continued belief in his existence.

Interviewer: But where is the truth? On the one hand, you seem to question whether there is a god. On the other hand, you believe in the existence of God. So that seems contradictory to me.
Preston: I don't believe in a god that is there for me, as in miracles and signs. I believe that there is an afterlife for me, that I won't go to hell and that I will be up there with the angels. That's what I believe in. But I am older now; I am not forced to go to church that much any more. So therefore I am starting to think now, "Do I really believe in the god that every Christian believes in?"

Whereas earlier in the interview, Preston has talked about his quest to have a sign from God, he now describes a different form of relation to God. He no longer believed in a personal god; he believed in the existence of an afterlife in heaven ("up there with the angels") rather than in hell. The text also articulated him as a person who, being older now, is in a position to make decisions about going to church, questioning his received beliefs. Yet the contradiction remains, for, as he says a little later, he feels "very gullible," and therefore subject to external forces. This Preston is, at one moment, an agent, and at another moment, a subject of influences.

Discussion

Reading interview transcripts in the way that I have done leads researchers away from a presentation of human beings in general and interviewees in terms of disembodied knowledge, conceptions, cognitive processes, and beliefs. Interview texts are recognized as the outcome of transactions involving interviewers and interviewees. Their specific utterances—produced in response to the previous utterance and to the entire history of utterances in the culture—are related to and intelligible through this activity. Because of the dialectical nature of an utterance, even (auto-) biographical narratives are not just about one person but also express persons in general. The narrator in situation is therefore different from the character in the plot (Roth, 2004). Thus, the interview transcripts are never only about Preston, they are also about how a religious person in general may experience the differences between science and religion. The interview is as much about ways of experiencing conflicting discourses as it is about one specific (concretely realized) way of experiencing them.

Choosing a narrative framework, where (auto-) biographical materials are not just about specific persons but also about types of persons, we do not have to ponder questions of the truth between what people say and what really happened to them, about their real beliefs and what they say they believe. We take any discourse as having a situated function; the relationship between any utterance and the overall activity is one of sense. The narrative framework allows us to see interview transcripts as establishing versions of the world, versions that have relevance in and pertain to the current situation. Different versions may be evident not only between different situations but also within a single situation, such as the same interview. Thus, all interviews with Preston and his classmates have to be seen in this perspective—participants were oriented to the production of an intelligible text about the nature of science, epistemology, learning, and religion. The interviewer and his interviewees are inherently responsible to one another for producing each meeting and intelligible text, and in doing so draw on culturally and historically available resources. As a result, both interview situation and interview text are concrete realizations of general possibilities—they are dialectical, constituting both particular instances and general cultural-historical possibilities.

Any attempt that seeks to evaluate the relation of truth between a ([auto-] biographical) text produced in the interview and some reality external to it (the interviewee's "true" life) is deemed to run into (philosophical, epistemological) trouble. No representation, as close to lived experience as it might lie, is already abstraction and not lived experience itself. There is an insurmountable (ontological) gap between what is and experience ("Erfahrung") and accounts of experience.

We may now begin to ask whether there is anything that we can do to assist students such as Preston in dealing with the conflict he experiences. Teaching premised on the goal to make students drop their religious commitments (e.g., accepting the creation narrative) seems to be undemocratic and inconsistent with the fundamentally human right to freedom of opinion and expression. Any attempt of undermining a person's belief system constitutes an act of (symbolic) violence, which questions not merely a way of thinking but in fact a way of being in the world. The purpose of education, therefore, has to lie in assisting students in understanding how others live, those who do not understand this difference are likely to experience science and religion as a conflict. In formal education, different teachers teach science and religion, often in different rooms. A clean split between the two domains is therefore achieved in spatial and temporal terms. Although our (ipse-) identities are a function of the situation, human lives are also characterized by a coherence of lived experience across situations. Students do not simply compartmentalize science and religion but in fact attempt to come to grips with both. Formal education has a role to play in assisting students navigate these situations where, as in Preston's case, science and religion cannot be compartmentalized but affect the person's entire being.

Teachers, individually and collectively, might want to think how they can provide opportunities not to convert students to one or the other discourse but of getting to know other ways of dealing with alternative discourses. At the present, schooling is not set up to teach for scientific literacy as a collective praxis, which, when implemented, would allow students to adhere to their individual discourses and at the same time teach them how to interact with others. Such an education would no longer focus on making everyone talk (think) in the same way about difficult (controversial) issues but allow students to develop competencies to participate in collective processes.

In the US, where there is a separation of church and state, a fundamental contradiction exists in the fact that schools, responsible for the education of future generation, are not allowed to deal head on with the issue. History shows that conflicts are then played out in legal and political arenas. In other countries (e.g., Austria, Canada, Germany), religion is taught in schools (even with students from multitudes of religious backgrounds) as a way of "educating the whole child [student]," and students take classes where they are provided with opportunities to understand the assumptions of other religions. Regarding the US, it appears ironic that in the same nation with one of the highest levels of education in the world, citizens cannot address the salient issues of conflicting discourses and life narratives.

Readers who find the argument difficult to accept and feel that Preston simply needed to be told differently forget that we all act in response to the constraints as we perceive and experience them. Many academics talk about how they are expected to sit on some committee or how the institution or others make them do something. Attempting to convince these academics otherwise will not help, because they continue to feel the expectation and the constraints on their actions, although others in a similar position may act very differently, and not experience such constraints. To understand and help Preston, if this is our goal, we need to start where he is at the moment, with his reality and the range of possibilities for action that he perceives.

Coda

Language has a central role in the situated constitution of different versions of the world; it inherently embodies possibilities for such versions; each version that is contingently produced is therefore a concrete realization of general possibility, and therefore is as much a social as an individual phenomenon. Language provides a number of resources that speakers can draw on in support of the different versions they constitute; these resources are also possibilities. The implications are clear. Any idea of eradicating ideas, beliefs, misconceptions, or versions of the world has to falter, given that the possibilities exist at the collective level in terms of linguistic means and resources of expression. Even if a student gives in to pressure (e.g., to do well on tests), and represses using certain resources, teachers will not have succeeded in eradicating the ideas, beliefs, misconceptions, or versions of the world.

Religion interacts with other aspects of life, which raises the potential for conflict. How we deal with these conflicts differs with culture, and there are no unique solutions available. In the US, students pledge allegiance referring to God, although the country's constitution enshrines a separation of church and the state. France, an equally secular state, legislated that students may not wear any evident signs of religion while in school, such as the hijab (Islamic veil), kippa (Jewish cap), or a cross. In Canada, Sikh policemen may wear the traditional turban, and women Islamic soldiers may cover their hair in a traditional way. Each society will need to work its way through the issues it associates with the relation between religion and other aspects of life. Language is the central medium with and through which this is achieved. Once we arrive at new ways to talk about the different domains at the collective level, we have also found new ways for individuals to account for the different domains in their own lives. There is, therefore, a need to engage in a collective endeavor for integrating religion, as there is a need for the individual to do the same.

References

Buber, M. (1970). *I and thou*. New York: Simon & Schuster.
Chinn, C. A., & Brewer, W. F. (1998). Theories of knowledge acquisition. In B. J. Fraser & K. G. Tobin (Eds.), *International handbook of science education* (pp. 97–113). Dordrecht, The Netherlands: Kluwer Academic Publishers.
Derrida, J. (1998). *Monolingualism of the Other; or, The prosthesis of origin*. Stanford, CA: Stanford University Press.
Latour, B. (1999). *Pandora's hope: Essays on the reality of science studies*. Cambridge, MA: Harvard University Press.
Lawson, A. E., & Worsnop, W. A. (1992). Learning about evolution and rejecting a belief in special creation: Effects of reflective reasoning skill, prior knowledge, prior belief and religious commitment. *Journal of Research in Science Teaching, 29*, 143–166.
Lynch, M., & Bogen, D. (1996). *The spectacle of history: Speech, text, and memory at the Iran-contra hearings*. Durham, NC: Duke University Press.
Mikhailov, F. (1980). *The riddle of self*. Moscow: Progress.
Ricœur, P. (1992). *Oneself as another*. Chicago: University of Chicago Press.
Rorty, R. (1989). *Contingency, irony, and solidarity*. Cambridge: Cambridge University Press.
Roth, W.-M. (2004). Autobiography as scientific text: A dialectical approach to the role of experience. *Forum Qualitative Sozialforschung / Forum: Qualitative Social Research, 5* (1). http://www.qualitative-research.net/fqs/fqs-eng.htm.
Roth, W.-M., & Alexander, T. (1997). The interaction of students' scientific and religious discourses: Two case studies. *International Journal of Science Education, 19*, 125–146.

Roth, W.-M., & Lucas, K. B. (1997). From "truth" to "invented reality": A discourse analysis of high school physics students' talk about scientific knowledge. *Journal of Research in Science Teaching, 34,* 145–179.

Wittgenstein, L. (1968). *Philosophical investigations* (3rd ed.). New York: Macmillan.

Examining the Evolutionary Heritage of Humans

David L. Haury

> *There is only one river.*
> *There is only one sea.*
> *And it flows through you, and it flows through me.*
> — Peter Yarrow[1]

If Galileo displaced us from the center of creation, has Darwin removed us from the top rung on the ladder of life? Does an evolutionary view of humans draw into question our uniqueness? Does acknowledging our biological heritage invalidate our religious roots? Are we merely modern descendants of apes?

Such questions may seem far removed from the practicalities of deciding what to teach in science classrooms, but perhaps the "modern culture wars" (Larson, 2004) do have their roots deep in the human psyche and our yearnings for meaning. Could it be that the mountains of evidence supporting evolution are not quieting the creationist opposition because they are missing the mark? The fact that evolutionary change is occurring and has brought us to our current condition in time seems indisputable if one simply acknowledges the reported findings from thousands of scholars from all parts of the world, for over a hundred years, from all cultures, and from a broad continuum of metaphysical beliefs. Many determined and noble-minded scholars go to extraordinary measures to document, point-by-point, why most arguments counter to prevailing scientific views of evolution do not hold up, but creationists seem unswayed by the preponderance of the evidence or the sophistication of the arguments.

Maybe it is not more evidence that is needed. Since childhood, I have been fascinated by each new report of human ancestors, and, until her passing, my conservatively religious mother would send me newspaper clippings that reported new anthropological finds, archeological digs, or new fossil species. When in college, I realized for the first time how many of my religiously minded friends did not welcome any news of non-biblical ancestors. Even more surprising to me, they questioned the sincerity of my own religious views because of my curiosity and fascination with our evolutionary heritage.

---

[1] This is a line from a folk song, *River of Jordan*, performed by Peter, Paul, and Mary.

The evidence for evolution has long been compelling for those who accept scientific evidence and reasoning, and new findings continue to fill in the details of the evolutionary epic. So, there will be no attempt in this chapter to make the case that human evolution occurs. Rather, the questions addressed here include practical matters for educators, such as: How do we best make the case for including human evolution in the school curriculum? What is the essential content? And what are the most constructive responses to those who oppose inclusion?

Making the Case for Studying Human Evolution
As idyllic as it would be to teach something simply because we find it interesting, overcrowded school curricula and the spread of high stakes testing render such a rationale inadequate. So we must think critically about an adequate justification for adjusting the life science curriculum to include treatment of human evolution. One advocacy group (Tennessee Darwin Coalition, 2001) has offered the following reasons:

- Leaving human evolution out of the curriculum renders inadequate the discussion of key ideas in biology, and prevents discussion of topics intrinsically interesting to students.
- The combination of a well-documented fossil record for human lineages and an abundance of evolutionary data derived from human DNA sequencing provides some of the best examples of evolutionary principles, so omitting attention to humans reduces the quality of the content that could be examined.
- Since much of the typical high school biology program focuses on human biology and human health issues, omitting attention to human evolution gives the false impression that principles of evolution have only limited application to us.

Unfortunately, there are no explicit recommendations to teach human evolution in most curriculum reform documents, such as the *National Science Education Standards* (NSES, National Research Council, 1996) in the US, or the *Benchmarks for Science Literacy* (AAAS, 1993). Similarly, most biology textbooks for schools in the US provide little or no coverage of the topic, and a recently published guide for teaching evolution (Bybee, 2004) makes no reference to humans, except for a passing reference to human genome information in one chapter. In short, national curriculum standards and major reform documents hold the door open to including human evolution in the school science curriculum, but offer little explicit support and provide little direction as to what should be covered.

The Context of Knowledge Domains
There are a variety of knowledge domains associated with any science topic, and the following domains provide rationales for learning about human evolution.

Evolutionary Theory
Consideration of evolution in general would be incomplete without studying the species that has had the most profound impact on the planet as a whole, and its diverse ecosystems. Klein (1999) asserted that the fossil record of the human family now provides one of the most compelling cases for macroevolution. In describing the major transitions in evolution, Maynard Smith and Szathmáry (1995) identify the transition from primate societies to human societies enabled by the advent of language as the most recent of eight major transitions in evolution since the appearance of replicating molecules. Wilson (2002) characterized one of our evolutionary attainments as becoming the first species in the history of the planet to constitute a geophysical force. In his words, "We have driven atmospheric carbon dioxide to the highest levels in at least two hundred thousand years, unbalanced the nitrogen cycle, and contributed to a global warming that will ultimately be bad news everywhere" (p. 23).

Nature of Science
Human evolution is a rapidly expanding field of science that offers a case study in how assumptions and generalizations based on insufficient data have been obstacles to fuller understanding. For instance, the traditional assumption of a single human lineage descending from primitive apes to our current condition cannot be supported by the evidence. There have been several human species, with some existing contemporaneously. For over a hundred years, our prior assumptions prevented us from realizing that we are not descendants of Neanderthals. DeSilva (2004) goes further:

> Perhaps the best topic teachers can use to exemplify the nature of science is paleoanthropology, the study of human evolution through the fossil record. Science educators have an opportunity to tackle "How do we know?" questions by examining evidences of our past and accurately defining the terms "hypothesis," "fact," "theory," and "belief." They can use recent discoveries to demonstrate that science is a self-correcting mechanism of understanding the world. By examining different hypotheses, they can encourage the skepticism, debate, and challenge to authority on which science thrives. (p. 257)

Diversity of Evidence
Synergistic lines of research from different domains of science also provide a unique glimpse of how different sciences make unique contributions to our understanding of a complex picture. A *Statement on Evolution and Creationism* by the American Anthropological Association (2000) illustrates the many lines of

research that have contributed to our understanding of human evolution, including: anthropology, archeology, comparative anatomy, genetics, geology, medical anthropology, medicine, paleoanthropology, paleoecology, paleoethnobotany, paleontology, primatology, and taphonomy.

Human Family
The study of human evolution provides a biological foundation for more completely understanding ourselves as biological beings. We learn the full extent to which we truly are part of a global family with a common heritage; we learn that we are all Africans, differing only in the paths taken by our ancestors since emerging as humans. Our traits, our diseases, and our pathways to health are all being shaped by evolution. It is in this context, I believe, where a deeper understanding of human evolution can provide the foundation for significant changes in perspective, leading to greater social justice. One lingering artifact of outdated views of human origins, indeed, creationist views of human origins, is the concept of "race." Many injustices have been overlooked, tolerated, or even encouraged in the past due to hierarchical thinking about human races, an ideology formulated in Europe. Not surprisingly, a "Caucasoid" skull discovered near the purported location of Noah's Ark and the supposed home of the original humans (Mukhopadhyay & Henze, 2003) was selected as representing the ideal form. Cultural norms, language, and prejudices emerging from this false notion have perpetuated deeply embedded social injustices, and it is time to set the record straight. We are all of one blood that began flowing in Africa, and the history of our migrations is recorded in our DNA (Sykes, 2001; Wells, 2002). This understanding alone is justification for teaching human evolution in school classrooms. What more important understanding could we possibly pass on to our children than the clear evidence that we are, indeed, biologically, one human family? It is not an ideal; it is not a euphemism; it is not positive thinking. It is a genetic reality, and it is evolutionary biology that enlightens us.

Ecological Identity
An understanding of human origins links us to a fuller understanding of our genetic connections in the tree of life; it explains our molecular and cellular relatedness to all life, a context for examining both the uniqueness of humans and the commonalities we share with other life forms. As Johanson (1996) has suggested, "This enlightenment about our place in nature we hope will enhance our sense of responsibility to the world around us." An examination of human evolution also forges a connection between life forces and the forces of nature expressed through climate and changing landforms. Humans emerged at a particular time and place and are, therefore, an expression of the interactions between environmental change and biological adaptation. We are the products of biological responses to a sequence of environmental conditions.

Worldview
An understanding of the scientific perspectives of human origins can complement one's religious beliefs. Though we would not presume to influence or study religious beliefs in a science classroom, how one thinks about humans in metaphysical terms can only be enhanced by clearly differentiating the biological dimensions of humans from the wider scope of reality posited by various metaphysical views.

Spirit of Discovery
Finding the pathway of human evolution amongst the complex array of molecular, fossil, and other findings is a wonderfully complicated challenge, filled with genuine human drama. One of my favorite stories is an account in *National Geographic* (Leakey, 1979) that describes how fossilized hominin[2] footprints were discovered as a researcher dived to the ground to avoid a missile of elephant dung flung by a colleague. For years, I enjoyed relating this story simply for the serendipity involved in this discovery, and the "down-to-earthness," so to speak, of field scientists. Later, I was delighted to learn that someone (WGBH, 2001, p. 24) used this event to engage students in studying what can be inferred from such trackways. Another sample activity that engages school students in this investigation has been presented by the Working Group on Teaching Evolution (1998, pp.81–86) of the National Academy of Sciences.

The Context of Design
One reason to examine human evolution in school biology courses is to present a reasoned, constructive response to those who wish to advance the idea of an intelligent designer. If there is a designer, why not start with ourselves and consider whether the evidence is more supportive of an intelligent designer or a biological process that continually reworks existing structures in response to environmental conditions? We could start as Ayala (2004a, 2004b) has suggested, with the human jaw. As he points out, our jaws are too small for the number of teeth we have, and any human engineer would have drawn up a better design. As it is, dentists and orthodontists make a pretty good living by pulling wisdom teeth to make more room, and realigning teeth. Does this seem

---

[2] The increase in our knowledge of primate evolution has recently led to a change in terminology for various groupings of related organisms. All Great Apes—including humans, chimpanzees, bonobos, gorillas, and orangutans—are now classified into one family, the Hominidae, or "hominids." In this scheme, at the tribe level between family and genus, modern humans, extinct human species, and all our immediate ancestors that walked upright, including members of the genera *Homo*, *Australopithecus*, *Paranthropus*, and *Ardipithecus*, are now known as the Hominini, or "hominins." The term "human" as used in this chapter refers both to modern humans and to all extinct species of the genus *Homo*. In short, the terms hominid, hominin, and human refer respectively to the family, tribe, and genus to which we belong in the current classification scheme.

the work of an intelligent designer, or the result of a remodeling process that led to a smaller jaw as brain size increased and diets changed over time?

Another example offered by Ayala is the narrow birth canal of humans that leads to thousands of deaths each year due to complications during delivery. Again, is this due to imperfections in design by a designer of questionable intelligence, or the natural consequence of a self-adjusting process responding to the dramatic increase in human head size? Ayala (in Freedberg, 2002) characterizes those who claim that a superior being designed human features as being guilty of blasphemy, saying, "If our organs have been designed by somebody, that person was very clumsy, outright stupid, and much worse than any human engineer." Though Ayala, a former priest, is very sympathetic to religious beliefs and acknowledges that science is not the only way of knowing, he demonstrates that we need look no father than our own bodies to find evidence for design by trial. On several occasions Ayala has made the crucial point that available evidence points to design without a designer. As designers, engineers start with raw materials and develop unique products or applications for particular purposes, but evolution can only modify the materials and structures that are already present. Both processes lead to an apparent design that is tailored to a particular set of conditions, but design solutions that are repeatedly constrained by modifications of previously existing materials seem more likely the result of an evolutionary process. So, perhaps the question is whether what seems to us like design requires a designer. Framing the question this way leads one to consider the design pathway rather than the remarkable outcome, and the telltale evidence of evolutionary "tinkering" seems abundant in the design pathways of biological adaptations. Another option, I suppose, is that we are at the mercy of a designer with the tendencies of Rube Goldberg.

Another strategy for comparing the reasonableness of intelligent design versus an evolutionary process is to consider the human genome. For example, there are over 2,000 documented cases of pseudo-genes (DNA sequences that have no apparent function) and nonfunctional long interspersed nucleotide elements (LINEs, thought to represent defunct retroviruses) in the human genome (Lander, Patrinos, & Morgan, 2001). Though it is possible that pseudo-genes and LINEs have some yet unknown functions, there seems no plausible explanation for them from a design point of view if they are as nonfunctional as they appear. Their existence is easily explained, however, from an evolutionary perspective; further, study of the conserved presence and positions of pseudo-genes provides information about biological relatedness and common ancestry (Goodman, 1999). Indeed, study of pseudo-genes allows the testing of predictions regarding presumed evolutionary relatedness that are based on other lines of evidence (Gregg, Janssen, & Bhattacharjee, 2003).

Is There a Case?

Human evolution is interesting to many of us, but there is no mandate from the science educational reform movement to teach it, and many school textbooks devote little or no attention to the topic. So, why include human evolution in the science curriculum? There are many discrete reasons that could be listed, but four seem particularly compelling: (a) some fundamental mechanisms and concepts of evolution are best illustrated or most well documented by findings related to the evolution of humans, so omitting attention to human evolution constrains the strength of lessons that could be learned; (b) almost everyone has questions about human origins and identity that can be addressed through study and discussion of human evolution, with an emphasis on clarifying our uniqueness as evolutionary beings, our evolutionary connections to the rest of nature, and the relatedness of all humans; (c) a study of human evolution provides an ideal context for examining the nature of science and scientific knowledge, with questions raised from religious grounds providing an ideal opportunity to compare and contrast various ways of knowing and their limitations; and (d) as we humans become more adept at using biotechnology to do our own evolutionary tinkering, it will become increasingly important that the general public make informed decisions regarding applications that could have evolutionary ramifications. In short, teaching about human evolution greatly enriches science content, it responds to fundamental curiosities and concerns that people have, it provides a personally meaningful context for examining scientific ways of knowing, and it informs decision making in an era of powerful biotechnologies.

The Essential Content

When human evolution is taught in schools, the emphasis is typically on the bones of ancestors and the inferred lineages among primates in general and hominids in particular. The changing line-up of human ancestors and relatives is certainly fascinating, but there is much more to the study of human evolution, and much of it relates to our own future and our emerging biotechnologies. An outline of possible dimensions to include in a basic study of human evolution might be:

Providing an Overview of the Emergence of Hominins
With the wealth of hominin fossils accumulating at a rapid rate, one could easily spend large amounts of time just becoming familiar with the various members of the human family tree. It seems less critical, though, that students know the names and locations of particular hominins than that they become familiar with some of the key transitions in human evolution, the types of evidence that are used to construct our mental picture of human evolution, and some of the active issues in the study of human evolution.

Converging Lines of Research
A striking feature of research in hominin evolution today is the confidence gained in interpreting findings when they coalesce from multiple, independent modes of investigation and yield complementary results. Making the case for migration patterns based on the distribution of fossils is greatly strengthened

when similar migration patters are inferred through DNA analysis. It seems particularly important that students learn about modes of analysis beyond recovery of fossils, as well as the principle that evidence of any sort gains credibility when it can be corroborated through a complementary, but independent, line of research. Given the increasing prominence of biotechnology in studying evolution, it is becoming increasingly important that students gain first hand experience with some of the tools of the trade, such as DNA analysis through electrophoresis, and analysis of genomic information. For instance, students can perform DNA extractions, DNA fingerprinting, or use DNA-DNA hybridization data to analyze genetic difference (Maier, 2004).

In recent years, there have also been many efforts to correlate phylogenetic changes in hominins, such as brain size, with other events, such as climatic changes, changes in diet, use of technology, and cultural evolution. For instance, three pivotal events seem to have occurred during the same period (Calvin, 2004) about 2.5 million years ago: stone toolmaking, the ice ages, and increased brain size. After that, the hominin brain steadily increased in size for the next 1.5 million years, and then there was another spurt about 750,000 years ago. What caused that? No one knows for sure, but perhaps it was related to the adoption of projectile hunting, or the advent of protolanguage (Calvin, 2004).

The important point to emphasize here is that evolutionary change is generally a response to some change in environmental conditions, so choosing a few examples to illustrate the point would be desirable. Transitions of particular importance to hominins, after attaining upright posture, include increases in brain size, changes in types of tools used, appearance of art, changes from roaming lifestyles to agriculture, and so on.

Diverging Branches on the Family Tree
The field of human evolution is one of rapidly expanding knowledge, from reports of astonishing new finds in the field by paleoanthropologists, such as finding possible *Homo habilis* specimens in Asia (Wong, 2003), to ever greater elucidation of our genetic heritage by molecular geneticists and molecular anthropologists. Understanding of hominin evolution is itself evolving at a rapid rate, and nothing could make that point with greater impact that the recent discovery of *Homo floresiensis* (Brown et al., 2004; Morwood et al., 2004) on the island of Flores in Indonesia. The astonishing discovery of a miniature form of human, that hunted pygmy elephants on a remote island until perhaps as recently as 13,000 years ago, reminds us of just how much more there is for us to learn about the family tree. This latest relative seems to have been a descendant of *Homo erectus*, as we are, and may have lived contemporaneously with modern humans for over 30,000 years. Since we are the only human species extant today, many are surprised to learn that in the past several species of humans were almost certainly alive at the same time. With this most recent discovery, for instance, it is likely that at least four species of humans were alive

at the same time within the past 100.000 years: *H. sapiens*, *H. floresiensis*, *H. erectus*, and *H. neanderthalensis*.

Though the discovery of *H. floresiensis* was unexpected, it fits with what we know of evolution. It has been clear for some time that the long-held notion of humans steadily progressing along a single pathway of evolution to our current highly evolved condition must give way. Just as the sudden appearance of *H. floresiensis* indicates, new forms arise through natural selection in the face of environmental changes, resulting in highly branched family trees. The evidence that *H. sapiens* and *H. floresiensis* could diverge so dramatically from the same ancestor, *H. erectus*, in such a short time span indicates how quickly divergence can occur. The finding also supports the notion that the diversity seen among modern humans reflects the impact of environmental conditions on local populations as humans migrated throughout the world from their place of origin in Africa.

Responding to Questions that People Have
Did we evolve from monkeys and apes?

It is evident from comments that people make that there is a general sense that we are most closely related to "monkeys and apes," and this characterization is often deliberately used to stir believers into action (Marks, 1995). It is unlikely that nonspecialists clearly differentiate monkeys from apes, so learning about human evolution could begin with examining the evidence that we are most closely related to chimpanzees. This would seem an opportune time both to highlight the various, complementary lines of evidence available, and the increased confidence in interpreting findings when the lines of evidence converge to form a composite whole. From the realm of paleoanthropology, bones recently discovered in Ethiopia (Woldegabriel *et al.*, 2001; Haile-Selassie, 2001) may be those of our earliest known ancestor living after the time when our evolutionary pathway branched from that of modern chimpanzees and bonobos. From the realm of molecular genetics, we have striking evidence of genetic similarities among humans, chimpanzees, bonobos, and other apes (see Figure 1). In terms of technology and behavior, chimpanzees are the only primate tool users other than humans to use the same tool on different objects, or to use two tools sequentially on the same object (Smith & Szathmáry, 1995), and for the benefit of those of us who teach, strategies like this are taught to offspring by mothers, with teaching accompanied by gestural communication (Boesch, 1993). This is the only known example of explicit pedagogy in the wild by nonhumans.

Does this mean that we are not particularly special in evolutionary terms?

This question is on the minds of many students and their parents, and thoughtful teachers can help people think constructively about this matter. It has long been popular for some (see for example, Morris, 1967) to emphasize that in a narrow sense, we are apes. Though this is true in a narrow sense, many bristle at the idea that we are not particularly special from an evolutionary point of view. Recently, much has been made of the pronouncement that the DNA of humans is 98% identical to the DNA of chimpanzees. Marks (2002) considers this a very overused "factoid" (p. 13), adding that it "bears the precision of modern technology; it carries the air of philosophical relevance." Further, "It reinforces what we already suspect, that genetics reveals deep truths about the human condition; that we are only half a step from the beast in our nature. Chimpanzees can be brutal, humans can be brutal; it's in our essence, it's in our genes." Marks does not question that we are genetically close to chimpanzees, but he believes that many have over-interpreted findings, as if there is self-evident meaning in the structural similarity of human and chimpanzee DNA. People need to understand, I believe, that even in evolutionary terms we are quite special, very successful, and not simply naked apes. Anyone who looks around them at the people in their offices and homes, and compares their behavior to the chimpanzees viewed in the wild through television can see the difference. Not just in terms of us having less hair and walking differently, but the profoundly different ways in which we live our lives and the ways we have fabricated our environments. Having students themselves come up with a list of distinguishing human features to examine may be a way of addressing the matter of uniqueness while giving students an opportunity to search for information. Figure 2 provides a sample of what such a list might include, noting that while not every listed feature is unique to humans, the array of features as a whole is unique.

We are not just smarter; our phenomenal ability to communicate through language and our ability to design sophisticated technologies has gained for us a powerful new mechanism of evolution: cultural evolution. We are not just biologically different; we have moved to a different evolutionary realm. Sure, we can identify many of the biological connections to our relatives and trace the roots of our emergence, and we need to attend to our physiological requirements as do all other life forms, but we are not just more of the same. The study of human evolution in school classrooms should openly celebrate the obvious: we are unique, there are evolutionary innovations that are uniquely expressed through humans. A book on my shelf (Hughes, 1999) provides insights into the evolutionary pathways that have resulted in remarkable feats of sensory perception among diverse organisms. The sensory abilities of many animals are quite extraordinary, and well beyond the capabilities of humans. The biggest difference between those evolutionary expressions and our own uniqueness, however, is that the accounts are recorded in books that we produced. Our taken-for-granted ability to reflect on evolution and think about how evolutionary pathways have given rise to our biological, technological, cultural, and even religious selves puts us in a unique position in the biological realm.

Does our knowledge of human evolution rule out the existence of a supreme being?

For most who resist learning about human evolution, it is the idea that human evolution rules out the existence of a supreme being that is probably the central concern. As Alters and Alters (2001) have pointed out, creationist leaders view the science classroom as a battleground where believers are subjected to a godless belief, evolution, that pulls the hearts and minds of young people away from God. Science teachers must respond to this heartfelt concern and should perhaps raise the question if students do not. After providing a straightforward answer, "no," this is an opportune time to make four points: (a) Although individual scientists may claim that evolution rules out for them the need for a creator, science as a way of knowing is unable to make any definitive claims about the presence or absence of a supernatural entity. There is no way to test the idea empirically. (b) A very high percentage of people believe in the existence of God, including some leading evolution scholars. Providing examples of evolution specialists who express a range of beliefs would be appropriate. (c) Evolution explains only our biological relationships to other life forms and provides explanations for our unique set of characteristics as biological beings. For those who believe there to be a nonmaterial or spiritual dimension to reality, examination of that dimension must come from elsewhere. (d) Science can make claims only about cause and effect relationships in the material world, so each individual must personally struggle with how best to situate scientific knowledge within a larger belief system. Studying human evolution can inform that process by providing answers to questions about our biological uniqueness and heritage.

Applying Our Knowledge of Human Evolution

Knowledge of human evolution has practical value; many of the issues of our day, particularly environmental issues and health issues, have evolutionary components. Wilson (2002, pp. 39–40) has eloquently made the point that the Earth is unlike any other known planet, because unlike other planets, the Earth is not in a physical equilibrium. "It depends on its living shell to create the special conditions on which life is sustainable. The soil, water, and atmosphere of its surface have evolved over hundreds of millions of years to their present condition by the activity of the biosphere, a stupendously complex layer of living creatures whose activities are locked together in precise but tenuous global cycles of energy and transformed organic matter." Within this cradle we evolved along with many other species, and when we alter the biosphere in any direction, we move the environment away from the "delicate dance of biology" and risk degrading the planet's great heritage and threatening our own existence.

The question is not whether the global environment will survive our impact— it will. Relevant questions that students can explore have to do with how we and other species will survive the environmental changes brought about by stresses we place on the Earth. Many believe we are in the early stages of the greatest mass extinction event since the end of the Cretaceous when dinosaurs became extinct, and the Species Survival Commission (2007) of the World Conservation Union maintains extensive documentation of endangered species.

Unlike mass extinctions of the past, this one is not caused by a cascading of catastrophic climatic events or a chance encounter with an asteroid, but is being caused by the environmental impact of an exploding human population and our unsustainable management of the Earth's resources. Not only are we related to other species through our evolutionary heritage, we are also interconnected to the physical environment through the ecological relationships that have evolved over time. That is a central principle of human evolution: we have evolved within a milieu of relationships, cycles, feedback systems, and energy transfers. Biological evolution responded to the checks and balances of nature, fitting organisms to environmental conditions, but our cultural evolution and technological evolution have enabled us to move beyond the natural constraints, for a time. Eventually, however, stressed systems will adjust or collapse, dramatic environmental changes will occur, and we will be facing new conditions to which we will have to adjust. The bright side of this scenario is that we are able to reflect on possible consequences due to our evolved intellectual capacity, and respond culturally and technologically. Knowledge of human evolution has given us both the warning and the opportunity to take corrective action.

In the realm of human health, an understanding of human evolution has given us powerful insights into the body's defense mechanisms, the challenges of infectious diseases, possible genetic therapies, and a wide range of potential cell-based treatments. The journal *Science* published a special issue on the Ecology and Evolution of Infection (11 May 2001) with an associated Web supplement, and an evolutionary analysis of human diets provokes thoughts about possible causes and responses to selected dietary problems (Ungar & Teaford, 2002). These are examples of the many potential issues and topics that could be examined and discussed in the context of human evolution. Spending time considering these issues both provides a dimension of personal relevance in studying human evolution and raises awareness of matters that have evolutionary roots as well as potential evolution-based solutions.

As an example, consider the case of sickle-cell anemia. Even though the disease may eventually kill its victims, the allele that causes sickle-cell anemia persists in populations originating from the Mediterranean basin, the Indian subcontinent, the Caribbean, portions of Africa, and parts of Brazil and Central America. On the surface, this does not make sense, but there is a well-known evolutionary explanation: those who are carriers of the sickle-cell allele but do not have the disease have a survival advantage in regions where malaria is present. This is an example of a human condition that seems counterintuitive, but an understanding of evolution explains the occurrence and provides guidance for responding culturally or technologically.

A Response to Those Who Oppose the Teaching of Human Evolution

First, help students see the grandeur. I believe it is important to celebrate the dignity of being an immensely successful evolutionary creature. Some attempt to counter creationist claims by downplaying our evolutionary uniqueness, while others suggest that we are an inevitable product of an anthropic principle (Morris, 2003). While there is much about being human that seems beyond the reach of science, what we are learning about our rise and ecological domination inspires awe of what life can accomplish through its evolutionary response to environmental conditions. Evolution exemplifies the potency of responding to challenges as opportunities, for that is seemingly just how evolutionary processes gave rise to our remarkable selves. As Andy Knoll of Harvard has remarked (Hayden, 2002), "The scientific narrative of the history of life is as exciting and imbued with mystery as any other telling of that story." There is, indeed, grandeur in the unique string of pivotal events that has formed us from a particular lineage of bipedal apes. During the last ten million years, our evolutionary response to changing ecological conditions and inhospitable environments has resulted in an enormous increase in brain size; a highly sophisticated ability to design and use tools; the development of self-awareness, language, and cultural adaptations; an ability to create our own environments and travel beyond our planetary abode; and the ability to be fascinated by our accomplishments.

Second, we must find a way to convey this view of grandeur without alienating students. As others (Smith, 1994; Smith & Scharmann, 1999) have so aptly noted, alienated students do not learn. Many students from religious homes or backgrounds hold views or beliefs that give rise to a wide range of feelings when confronted with human evolution, from discomfort to antagonism. We must openly acknowledge this range of attitudes and feelings, and respond in some way that does not alienate students before engaging them in an examination of our evolutionary heritage.

One approach is that of Gould (1999) who speaks of respectful non-interference, with science and religion (or worldviews) being non-overlapping knowledge domains that by their very nature cannot be in direct conflict. One possible implication of this idea is that a teacher should acknowledge these two knowledge domains, assure students there will be no attempt to challenge or change anyone's worldview or religious position, and get on with examining what science has to say about human heritage. This seems a very constructive approach, but perhaps too idealistic to be practical. As Gould himself noted (p. 148), one problem is that this line of epistemological demarcation does not neatly address most problems that arise in the classroom because they are not purely conflicts between the knowledge domains of science and religion. Almost all scientists and most major religious leaders are on the same side of the evolution issue, opposing the views of creationists, and not all creationists hold strongly religious views. This is what significantly complicates classroom challenges; it is not at its core a straightforward religion versus science issue.

One approach through this complicated issue would be to simply throw open the doors to all knowledge domains and structure the learning about human evolution around questions that students have, leaving open to consideration all viewpoints that students bring to the forum. I actually tried this approach many years ago while teaching in South Australia. The Australian Science Education Project (ASEP) had developed a trial edition of a science unit entitled "Where Humans Came From," and like all ASEP units, learning was largely self-directed, with the teacher having the role of facilitator. The introduction to the unit offered some possible alternatives to consider, including: creation, spontaneous generation, evolution, or influence of beings from another world. In this unit, students kept diaries of their examinations of various documents provided, including religious writings, and they produced individual reports before engaging in a classroom symposium. Truthfully, my students enjoyed this experience, a wide range of views were expressed, and few students seemed threatened by the experience. There were several aspects of this instructional experience, though, that concerned me. Most students used the occasion to reify what they already believed coming into the situation; it was largely a "reading, writing, and talking" experience; and it yielded much class time to examination of materials that were clearly not in the domain of scientific knowledge. This approach, though welcoming of all views, seemed on the whole ineffective in addressing the essential science content and moving students toward an understanding and appreciation of our evolutionary heritage.

Is there a non-threatening, non-alienating middle ground that enables us to engage students in considering the emerging picture of our own evolutionary heritage? There has to be; human evolution is too important a topic to leave unexamined. Dagher and BouJaoude (1997) found, for instance, that "teaching students about the nature of scientific facts, theories, and evidence is more likely to enhance understanding of evolutionary theory if students are given the opportunity to discuss their values and beliefs in relation to scientific knowledge." Certainly an effective instructional approach would openly acknowledge the seemingly conflicting perspectives of science and some worldviews, using the occasion to discuss the nature of science and distinguish scientific ways of knowing from other knowledge domains. Further, it seems worth acknowledging the intelligent design arguments and taking the opportunity to make two points: (a) intelligent design is not science because it is not testable, and (b) if it were to be subjected to scientific scrutiny, it would fail the test of logic because the evidence from human evolution does not support the idea. There are very specific examples in our evolutionary pathway that bring into question the wisdom of a supposed supernatural engineer. Though the available evidence does not seem commensurate with a designer who periodically intervenes in nature to create new life forms, that doesn't mean that the highly complex, evolving world that is our cradle and home was not created. If there is an intelligent creator, that designer produced processes as well as materials, and our scientific minds ponder the inner workings of these evolutionary processes.

An effective approach to teaching human evolution must also be pedagogically sound in responding to students' needs and interests, engaging them in active learning by asking questions, examining data, drawing conclusions, and making connections with daily life. Students should come to see that there is more to a study of human evolution than examining a magnificent parade of ancestral bones, and that our understanding of human evolution is not just about explaining our origins. It is about what and who we are, biologically, and the implications of that awareness for how we take care of ourselves, how we treat each other, and how we impact our evolutionary homelands. We are still evolving, but the significant difference between us and the rest of the biological world is that we have moved far beyond biological evolution to a condition where biological evolution, cultural evolution, and technological evolution are interwoven, and our enormously large brains enable us to use scientific findings, personal knowledge, and collective value systems to make decisions about all three strands of our evolutionary pathway. To some degree, we can become what we think. That is unique to us, and it seems important that more of us come to realize the importance of that possibility, or should we say, responsibility. As others (Lemonick & Dorfman, 1999) have so forcefully put it, "After millions of years, evolution by natural selection, operating blindly and randomly, has produced a creature capable of overturning evolution itself. Where we go from here is up to us." Perhaps an overstatement, but the point is well taken: we humans have progressed well beyond finding solid evidence for human evolution to having the capacity to harness evolutionary mechanisms for our own purposes. This is another strong reason for public understanding of human evolution; not to explain the past, but to make informed decisions about our evolutionary future.

Finally, in the special case of humans, we should distinguish between humans as biological beings with an evolutionary relationship to the rest of nature, and the belief of many that "humanness" includes a spiritual dimension. If, on one hand, we claim that scientific explanations cannot incorporate the notion of a supernatural entity, then we must also acknowledge that science is incapable of determining one way or the other if there is a nonmaterial dimension to human life. Science seeks material explanations for events and processes in a material world, but can make no claims for any nonmaterial dimensions of reality that may exist. This humble acknowledgement should be communicated to students, and there seems no more powerful context in which to position that acknowledgement than within the context of an evolutionary view of human biology. For instance, though we can confidently say that science provides abundant evidence that humans have an evolutionary heritage, science cannot say with authority that we are *just* an accidental by-product of chance events with no special claim to favor by a supreme being. Some may believe that, and they may use the theory of evolution to bolster their claims for there being no creator, but such a notion cannot be asserted as a scientifically determined reality. Indeed, it has been suggested (Goodenough, 1998) that the scientific account of nature could be called the Epic of Evolution, and that this is the one story that actually has the potential to unite us because it happens to be true. Contributing to the realization of that potential seems a worthy goal for any science teacher.

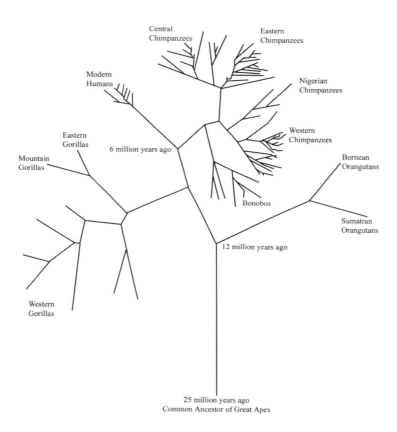

Figure 9.1 Genetic diversity among the great apes. Adapted from Zimmer, 2001.

Biological features

Brain/intellect
Hand
Bipedal posture (appears by at least about 4 mya; footprints)

Behavioral features

Language
Fabricate clothing and shelter
Process food
Domesticate animals

Social and Cultural features

Development of cities and states
Government
Migrations of humans begin with *H. erectus* (about 1.7 mya) (worldwide species, not confined to a particular habitat)
Colonization
Emergence of farming
Expression through arts

Technological features

Tools (appear about 2.5 mya; archeology begins)
Fire

Conscience

Sense of sacredness
Awareness of Earth
Self-consciousness
Morals

Figure 9.2 Selected features that differentiate humans from other species.

References

Alters, B. J., & Alters, S. M. (2001). *Defending evolution in the classroom: A guide to the creation/evolution controversy.* Sudbury, MA: Jones and Bartlett.
American Anthropological Association. (2000). *Statement on evolution and creationism.* Available online at http://www.aaanet.org/stmts/evolution.htm.

American Association for the Advancement of Science. (1993). *Benchmarks for science literacy*. New York: Oxford University Press.

Ayala, F. J. (2004a). Arguing for evolution. In R. W. Bybee, *Evolution in perspective: The science teacher's compendium*, pp. 1–4. Arlington, VA: NSTA Press.

Ayala, F. J. (2004b). Design without designer: Darwin's greatest discovery. In W. A. Dembski & M. Ruse, *Debating design: From Darwin to DNA*. Cambridge: Cambridge University Press.

Boesch, C. (1993). Aspects of transmission of tool-use in wild chimpanzees. In K. R. Gibson & T. Ingold (Eds.), *Tools, language and cognition in human evolution*, pp. 171–83. Cambridge: Cambridge University Press.

Bybee, R. W. (Ed.). (2004). *Evolution in perspective: The science teacher's compendium*. Arlington, VA: NSTA Press.

Calvin, W. H. (2004). *A brief history of the mind: From apes to intellect and beyond*. Oxford: Oxford University Press.

Dagher, Z. R., & BouJaoude, S. (1997). Scientific views and religious beliefs of college students: The case of biological evolution. *Journal of Research in Science Teaching, 34*, 429–445.

Darwin, C. R. (1859). *On the origin of species by natural selection*. London: J. Murray.

DeSilva, J. (2004). Interpreting evidence: An approach to teaching human evolution in the classroom. *The American Biology Teacher, 66* (4), 257–267.

Freedberg, L. (2002). Intelligent design's public defender. *San Francisco Chronicle*, p. D-1.

Goodenough, U. (1998). *The sacred depths of nature*. New York: Oxford University Press.

Goodman, M. (1999). The genomic record of humankind's evolutionary roots. *American Journal of Human Genetics, 64*, 31–39.

Gould, S. J. (1999). *Rocks of ages: Science and religion in the fullness of life*. New York: Ballantine.

Gregg, T. G., Janssen, G. R., & Bhattacharjee, J. K. (2003). A teaching guide to evolution: Discovering evolution through molecular evidence. *The Science Teacher, 70* (8), 24–31.

Haile-Selassie, Y. (2001). Late Miocene hominids from the Middle Awash, Ethiopia. *Nature, 412*, 178–181.

Hayden, T. (2002). A theory evolves: How evolution really works, and why it matters more than ever. *U.S. News & World Report*, 44–50.

Hughes, H. C. (1999). *Sensory exotica: A world beyond human experience*. Cambridge, MA: MIT Press.

Johanson, D. C. (1996). Human origins. *National Forum, 76* (1), 24–27.

Klein, R. G. (1999). *The human career: Human biological and cultural origins*, second edition. Chicago: The University of Chicago Press.

Lander, E.S., Patrinos, A., & Morgan, J.J. (2001). Initial sequencing and analysis of the human genome. *Nature* 409, 813–958.

Larson, E. J. (2004). *Evolution: The remarkable history of a scientific theory*. New York: Random House.

Leakey, M. (1979), Footprints in the Ashes of Time, *National Geographic*, 155, 446–457.

Lemonick, M. D., & Dorfman, A. (1999, August 23). Up from the apes. *Time, 154* (8), 50-58.

Maier, C. A. (2004). Building a phylogenetic tree of the human & ape superfamily using DNA-DNA hybridization data. *The American Biology Teacher, 66* (8), 560–566.

Marks, J. (2002). *What it means to be 98% chimpanzee: Apes, people, and their genes*. Berkeley: University of California Press.

Marks, P. (1995). *Someone's making a monkey out of you!* Colorado Springs: Master Books.

Morris, D. (1967). *The naked ape: A zoologist's study of the human animal*. New York: Random House.

Morris, S. C. (2003). *Life's solution: Inevitable humans in a lonely universe*. Cambridge: Cambridge University Press.

Morwood, M. J., Soejono, R. P., Roberts, R. G., Sutikna, T., Turney, C. S. M., Westaway, K. E., Rink, W. J., Zhao, J. –X., Van Den Bergh, G. D., Due, R. A., Hobbs, D. R., Moore, M. W., Bird, M. I., & Fifield, L. K. (2004). Archaeology and age of a new hominin from Flores in eastern Indonesia. *Nature, 431*, 1087–1091.

Mukhopadhyay, C., & Henze, R. C. (2003). How real is race? Using anthropology to make sense of human diversity. *Phi Delta Kappan, 84* (9), 669–678.

National Research Council. (1996). *National science education standards*. Washington, DC: National Academy Press.
Smith, M. U. (1994). Counterpoint: Belief, understanding, and the teaching of evolution. *Journal Research in Science Teaching, 31*, 591–597.
Smith, M. U., & Scharmann, L. C. (1999). Defining versus describing the nature of science: A pragmatic analysis for classroom teachers and science educators. *Science Education, 83*, 493–509.
Maynard Smith, J., & Szathmáry, E. (1995). *The major transitions in evolution*. Oxford: W. H. Freeman.
Species Survival Commission. (n.d.). Retrieved June 7, 2007, from http://www.iucn.org/themes/ssc/
Sykes, B. (2001). The seven daughters of Eve: The science that reveals our genetic ancestry. New York: W. W. Norton.
Tennessee Darwin Coalition. (2001). Position statement: Teaching human evolution in the high-school classroom. *Reports of the National Center for Science Education, 21* (1–2), 5–6.
Ungar, P. S., & Teaford, M. F. (2002). *Human diet: It's origin and evolution*. Westport, CT: Bergin & Garvey.
WGBH/Nova Science Unit & Clear Blue Sky Productions. (2001). *Teacher's Guide. Evolution: A journey into where we're from and where we're going*. Boston, MA: author.
Wells, S. (2002). *The journey of man: A genetic odyssey*. New York: Random House.
Wilson, E. O. (2002). *The future of life*. New York: Knopf.
Woldegabriel, G., Haile-Selassie, Y., Renne, P. R., Hart, W. K., Ambrose, S. H., Asfaw, B., Geiken, G., & White, T. (2001). Geology and paleontology of the Late Miocene Middle Awash valley, Afar rift, Ethiopia. *Nature, 412*, 175–178.
Wong, K. (2003, November). Stranger in a new land. *Scientific American, 289* (5), 74–83.
Working Group on Teaching Evolution. (1998). *Teaching about evolution and the nature of science*. Washington, DC: National Academy of Sciences.
Zimmer, C. (2001). *Evolution: The Triumph of an Idea*. New York: Harper Collins.

Approaching the Conflict between Religion and Evolution[1]

Lee Meadows

Biology teachers often approach their students' religious objections to evolution with a resolution mindset. They may take it as their responsibility to tell their students how to resolve the conflict, they may try to help the students figure out ways to resolve the conflict, or they may simply teach as if the conflict is already resolved, expecting religious students to leave their conflict at the door before entering the classroom. They fall into the resolution trap; Stephen Jay Gould (1999) warns both the scientific and the religious person to beware:

> When we must make sense of the relationship between two disparate subjects (science and religion in this case)—especially when both seem to raise similar question at the core of our most vital concerns about life and meaning—we assume that one of the two extreme solutions must apply: either science and religion must battle to the death, with one victorious and the other defeated; or else they must represent the same quest and can therefore be fully and smoothly integrated into one grand synthesis (p. 51).

Resolution is ineffective; "two disparate subjects" cannot be resolved. Gould instead calls for "a 'golden mean' that grants dignity and distinction to *each* subject" (p. 51, emphasis in the original). In the case of teaching biology to religious students, Gould's golden mean is a focus on conflict management (Meadows, Doster, & Jackson, 2000), not conflict resolution.

In this chapter, I present conflict management as a way to help students study evolution when their religious beliefs create conflict with evolution. After setting the stage with a case study of a student feeling the conflict between her faith and science, I present the research leading to a management approach and management principles biology teachers can use in their classrooms. For simplicity's sake, I focus only on biological evolution, but these thoughts should translate to the teaching of evolution in other science fields, such as the teaching of stellar evolution (Shipman *et al.*, 2002; Brickhouse *et al.*, 2000) or geological evolution.

---

[1] I'm indebted to David Hannych for encouragement and insight on early drafts.

Joanne's Story

Joanne is a ninth grader growing up in north Mississippi in the USA. She likes the small town she lives in—her family and school are both good, and her faith gives her life a deep sense of meaning. When she encounters hard times in her life, as when her favorite aunt died suddenly, faith gives her a lot of comfort, and she knows that God listens to her prayers. She's worried, though, because she's starting ninth grade biology, and toward the end of the course, she'll have to study evolution.

Joanne has dreaded the evolution unit for several years now. Science has always been her favorite subject in school, but she's often felt tension between her faith and her love for science. Sometimes, she even feels guilty. She goes to a Baptist church where she's learned that the Bible should be read literally. She's never questioned that teaching; everyone in her life believes basically the same way. For Joanne, if the Bible says that God created the world in six days, that's what happened. In fact, as she imagines the creation of the world, she sees God's majesty and power at work as he shapes the world like a master designer.

Joanne is just beginning to realize that her love for creation is probably the basis for her love for science. She sees science as the study of God's world, and everywhere she looks, she sees evidence of God at work. When she walks in the pine forests next to her house, she finds herself at times worshipping God, especially if a rare Mississippi wind has the treetops swaying in unison. On an elementary school field trip to a hands-on science museum in Memphis, she first realized how much she liked science. As she explored the exhibits, she realized that God lets people figure things out about the world through scientific experiments. Now, she feels like she's uncovering God's truth when she studies science. Last year, her middle school teacher was part of a special project from the university nearby, and Joanne got to spend nine weeks using a new, hands-on chemistry curriculum. She saw God there, too, in the midst of the beakers, chemicals, and balances. As she began to really understand the intricacies of molecules and atoms, she was amazed that God could think up a universe designed on such small building blocks.

As she thinks about taking biology, though, she's worried. She's heard from older students in her church youth group how bad the evolution unit is. One of the sophomores talked about an argument on creation versus evolution in biology class. He said that the teacher got mad and finally settled the argument by telling the class, "You have to choose between science and faith. You can't have both." After hearing that at youth group, she asked her youth pastor about some of her worries. He said, "You just have to believe, Joanne. You can't think all this through. God wants us to believe in him, not look for proof, not in science or anywhere else." Joanne felt ashamed at that moment for how much time she spent thinking about how science explains how the world works. Joanne thought about going to her pastor and asking him some of her questions, but she's a little scared of him. One time from the pulpit, he talked about how scientists, especially "evolutionists", are part of the Devil's agenda because they want to stamp out people's faith. She has now decided that she'd better keep her love of science to herself, especially when she is at church.

Deep down, Joanne's biggest worry is that the school biology teacher really is right and that she does have to choose between her science and her faith. Up until now, science and faith have both made her life better. She sees beauty and truth in both, and both show her meaning in the world around her. What will she do, though, if evolution and the Bible really do conflict? The book of Genesis says that God created the world in six days; science says it took billions of years. If God had meant he created the world over a long period of time, wouldn't he have just said that? If evolution is correct, she's scared that she might lose almost everything that's precious to her. She won't have a Savior, eternal life, a church family, prayer, Scripture, or even God. The stakes are very high in Joanne's mind, and she really wishes that her teacher would just skip over evolution all together.

Clashing Worldviews
High school biology teachers have religious students in their classes. Students from Christian, Jewish, Muslim, Native American, and other faiths can have belief systems that conflict with evolution. The conflict can be small, creating minor apprehensions, or large, creating distress and resistance to studying evolution like in the case of Joanne. Children raised in religious traditions emphasizing literal readings of their respective holy writings will probably see the most conflict between their faith and evolution. Children from other religious traditions, even if they haven't been taught literal understandings of their scriptures, still may see conflict in the evolution classroom. For example, students raised in liberal Christian or Jewish traditions may feel a subtle pressure to reinterpret their faith so that it matches scientific explanations.

Some members of the scientific community perpetuate the conflict religious students feel with science by attacking religion, as Miller (1999) eloquently describes:

> [There is a] reflexive hostility of so many within the scientific community to the goals, the achievements, and most especially to the culture of religion itself. This hostility ... sharpens the distinctions between religious and scientific cultures, produces an air of conflict between them, and dramatically increases the emotional attractiveness of a large number of anti-scientific ideas, including creationism (pp. 166–167).
>
> The backlash to evolution is a natural reaction to the ways in which evolution's most eloquent advocates have handled Darwin's great idea, distilling from the raw materials of biology an acid of hostility to anything and everything spiritual (p. 189).

Miller also describes the reaction from religious people when Darwin's theory is stretched out of its proper scientific boundaries and used as a grand unification theory for social problems to which evolution can't speak. This broadening of evolution can directly conflict with religious mores.

> Less than half of the US public believes that humans evolved from an earlier species. The reason ... is because of well-founded belief that the concept of evolution is used routinely ... to justify and advance a philosophical worldview that they regard as hostile and even alien to their lives and values (p. 167).
>
> Little wonder that people who see the world as a place of deliberate moral choice, who see clear differences between good and evil, and who cherish virtues such as courage, honesty, and truthfulness would take issue with [materialistic evolution]. Since such characterizations are presented as the direct implication of evolutionary theory, one might fairly conclude from them that evolution itself is their enemy (p. 171).

Certainly not all or even most scientist have such an intolerant view of religion, but those who do express this view perpetuate the conflict religious students feel as they venture into the science classroom.

This conflict results from the continuing clash of two different worldviews, one religious and one scientific. Religion and science have differing sets of assumptions about how the world works and differing rules about what constitutes truth. They're inherently different ways of knowing the world, and they can't and shouldn't be fused together. Gould (1999) concurs:

> People of goodwill wish to see science and religion at peace, working together to enrich our practical and ethical lives. From this worthy premise, people often draw the wrong inference that joint action implies common methodology and subject matter—in other words, that some grand intellectual structure will bring science and religion into unity (p. 4).

Exacerbating the conflict is the prior assumptions with which people approach the conflict. Some scientists hostile to religion hold a presupposition of scientific materialism. Miller (1999), speaking of Harvard geneticists Richard Lewontin, states, "Lewontin's eloquent hostility to anything connected to God rises naturally from his conception of science as the 'only begetter of truth'—a conception, incidentally, that relies on a prior commitment to philosophical materialism" (p. 186). Philosopher Alvin Plantinga (2000) describes the origin of these differing worldviews this way:

> You may think humankind is created by God in the image of God—and created with a natural tendency to see God's hand in the world about us ... Then, of course, you will not think of belief in God as ... a manifestation of any kind of intellectual defect ... It is instead a cognitive mechanism whereby we are put in touch with part of reality—indeed, by far the most important part of reality ... On the other hand, you may think we human beings are the product of blind evolutionary forces; you may think there is no God and that we are part of a godless universe. Then you will be inclined to accept the sort of view according to which belief in God is an illusion of some sort, properly traced to wishful thinking (pp. 190–191).

Naturally, Plantinga is probably describing poles: at one end is the religious thinker with no scientific presuppositions; at the other end is a scientific thinker with no religious presuppositions. People can be in between the two poles.

Worldview clashes can turn the biology classroom into a staging ground for deep conflict, and biology teachers can even perpetuate the warfare. This is especially true when biology teachers approach the conflict with a resolution mentality, taking as theirs the responsibility to cause students to resolve the conflict between the two clashing worldviews. The warfare can be subtler, though. Some biology teachers, in surrender to the conflict, leave evolution completely out of the curriculum (Jackson *et al.*, 1995). Others may skirt the conflict by focusing simply on scientific evidence and avoiding discussions of any of the larger implications of evolution. A more effective approach is conflict management, an approach growing out of research conducted with science teachers who themselves were from literalistic religious traditions.

Research Background

The conflict management approach to the teaching of evolution grew out of my collaboration with David Jackson, a science educator at the University of Georgia. Our first work together, "Hearts and Minds in the Science Classroom: The Education of a Confirmed Evolutionist" (Jackson *et al.*, 1995), chronicles David's growth as he learned how a different set of life experiences can deeply impact science teachers' approaches to evolution in the classroom. David, an agnostic, had never worked with science teachers who also held to a deep faith until he moved to Georgia in the USA. David was surprised to find some science teachers who were staunchly opposed to teaching evolution in their classes. At first, David tried to correct their beliefs about evolution, but then he began to realize that he had skipped the essential first step of listening to them before trying to influence them. He began to find that, rather than being uninformed, many of these teachers were thinking through their religious beliefs, their scientific beliefs, and the interplay between the two. He began to see that science teachers had to consider the hearts, as well as the minds, of their students. Many of the teachers in the study, and by extension religious students like them in science classes, are actively choosing not to learn about evolution. They place more value on the heart decisions of their faith, as Joanne's story above illustrates. Evolutionary science pales in importance to the eternal issues of God, Heaven, and salvation.

I know well this tension between the heart and the mind because I've lived it. I was raised in a Christian fundamentalist home and church, and I'm now a science teacher and educator. Working through this tension was a perspective I brought to the Hearts and Minds study. My own faith journey has led me away from fundamentalism, but I do still hold to the view that the Christian scriptures are the inspired words of God. I find truth in both worldviews. Science provides truth from the basis of evidence, but my faith also provides an intellectual, durable system of knowing the world. Plantinga (2000) argues for the reliability of faith using the following metaphor:

> Faith ... is far indeed from being a blind leap; it isn't even remotely a leap in the dark. Suppose you are descending a glacier at twelve thousand feet on Mount Rainier; there is a nasty whiteout and you can't see more than four feet before you. It's getting very late,

the wind is rising and the temperature dropping, and you won't survive ... unless you get down before nightfall. So you decide to try to leap the crevasse before you, even though you can't see its other side and haven't the faintest idea how far it is across it. *That's* a leap in the dark. In the case of faith, however, things are wholly different. You might as well claim that a memory belief, or the belief that 3 + 1= 4, is a leap in the dark. What makes something a leap in the dark is that the leaper doesn't know and has no firm beliefs about what there is out there in the dark—you might succeed in jumping the crevasse and triumphantly continue your descent, but for all you know you might instead plummet two hundred feet in to the depths of the glacier. You don't really *believe* that you can jump the crevasse ... you *hope* you can, and act on what you *do* believe—namely, that if you don't jump it, you don't have a chance (p. 263).

These perspectives set up an odd tension. Christian friends ask "How can you be scientist and still be a Christian?" Science friends ask "How can you be a Christian and still be a scientist?" I'm comfortable now with the conflict between my dual worlds because I've learned to manage, not resolve, the conflict between the two.

I envision a biology classroom where religious students are invited into the study of evolution without threat to their religious beliefs. Creating that kind of classroom requires that teachers help their religious student to manage, not resolve, the conflict (Meadows, Doster, & Jackson, 2000). The idea of conflict management grows from three key assumptions: religious students who have been taught a literal interpretation of their scripture's accounts of origins will almost always see a deep conflict between their faith and evolution, conflict between a literal religious view of origins and scientific evolution are deep and pervasive, and biology teachers have no place questioning students' religious beliefs. In other words, the biology classroom is not the place to attempt to iron out the conflict between evolution and religion. Instead, teachers can assist religious students in managing the conflict so that those students can engage in the study of evolution as much as possible.

Management Approaches

Biology teachers can do several things to assist students with learning about evolution when their students' religious beliefs conflict with evolution. I offer below five guiding principles for inviting religious students into the study of evolution through a management approach. These principles target the teaching of students who take a literal approach to their religion's creation accounts, but I'll also address how to modify these guidelines for students from non-literalist backgrounds. I focus on literalist students because they are most apt to experience conflict between their religious and scientific beliefs.

Respect your students' religious belief

Good science teachers know the value of showing respect to their students. Respect can often solve classroom management battles, diffuse conflict with parents, and win over students who enter the classroom with bad attitudes about biology or even school in general. Biology teachers should extend this same type of respect to students who come from religious traditions that create conflict with the study of evolution. They should not try to remove their students' religious beliefs, especially not from a perspective that science is superior to religion. As Gould (1999) says, "If religion can no longer dictate the nature of factual conclusions residing properly within the magisterium [i.e., worldview] of science, then scientists cannot claim higher insight into moral truth from any superior knowledge of the world's empirical constitution" (pp. 9–10). Science can't trump religion in the worldview clash.

Teachers don't have to agree with their students' religions to show them respect. They can simply offer students the opportunity to express their apprehensions by means of a question such as "You seem to be bothered by this topic. What are you thinking?" Teachers can offer to listen to students' concerns either during or after class. If teachers communicate that they are genuinely concerned about students and their journey of learning, even if not religious themselves, they'll do much to create a safe environment for students to explore the tension-filled territory of the intersection between science and their faith.

The opposite of this type of respect is trying to undermine or eradicate students' religious beliefs. Just as teachers in a public school have no right to advocate religion, they have no right to tear down students' religious beliefs. As parents of two school-age children, my wife and I take very seriously our responsibility to raise our sons in our faith. We will contend sharply with any teacher who crosses our line of authority and undermines the faith we are teaching our sons. For example, we would take issue with a biology teacher who denigrated the value of faith following the advice of Good (2003) to readers of *The American Biology Teacher*: "When young children are indoctrinated into believing that for which there is no evidence (God, Heaven, Hell, etc.), a habit of mind is being developed that is inconsistent with the open, inquiring mind needed for scientific study" (p. 515). Faith is not necessarily inconsistent with open-minded inquiry; Good is perpetuating a stereotype of religion as close-minded. Consider also that we're in an era of *Science for All Americans* (Rutherford & Algren, 1990) when we need to invite all students into the study of science. Most biology teachers understand the need to engage girls, minorities, and other students typically underrepresented in the science classroom. We need to extend this same level of respect to students from religious traditions that might oppose learning about evolution. Are we going to celebrate diversity in all its shades, including religious traditions, or are we going to require students to fit a standard science image before engaging them?

Present Evolution as an Undeniable Scientific Understanding
Although biology teachers need to respect the beliefs of religious students, they need also to attempt to gently and firmly guide students to see that whether or not evolution has occurred is not debated among scientists. Evolution is treated

as fact by the scientific community; creationism has no place in the biology classroom. As Miller (1999) has clearly argued:

> In the real world of science, in the hard-bitten realities of lab bench and field station, the intellectual triumph of Darwin's great idea is total. The paradigm of evolution succeeds every day as a hardworking theory that explains new data and new ideas from scores of fields. High-minded scholarship may treat evolution ... as just another scientific idea that could someday be rejected on the basis of new data, but actual workers in the scientific enterprise have no such hesitation—they know that evolution works. It works as a continuing process, and it works as a historical framework that explains both past and present (p. 165).

Religious students do not need to encounter a softened version of evolution in the high school classroom and later find out that astronomers, geologists, and biologists go about their work with the presumption that the universe, the earth, and life on earth all are the product of evolution. Details of the mechanisms of evolution are still under debate by scientists, but that it occurred is not.

Here is where biology teachers may find non-literalist students struggling with conflict, also. Students from liberal traditions may have been taught theistic evolution, the idea that God created the world through evolution. This belief may be a management strategy, helping students to make sense of the interplay of religion and science in the big questions of origins, but it is unscientific. Scientific conclusion must rest on natural evidence only. Scientific explanations of the origins of life cannot invoke the actions of God or any supernatural act.

This recommendation, when combined with the first on respect, sets up a tension for biology teachers. How can they gently show religious students the scientific understanding of evolution, knowing that doing so will probably cause conflict for many students? First, biology teachers should approach this process with respect. Teachers should recognize and even tell their students that they know that this study is difficult. Teachers should accommodate the pace of lessons to allow students time and energy to process some of this conflict, expecting some juvenile emotional reactions. These students are teenagers, not adults. Teenagers often don't handle deep conflict rationally and calmly. Second, teachers should let the evidence speak for itself. If teachers start preaching evolution to the students, students will want to preach their religious gospels back. Instead, teachers should set up lessons where students examine the evidence for evolution first hand. Following the principles of good inquiry teaching (National Research Council, 2000), teachers need to guide learners to explanations via a "priority to evidence" (p. 25) so that "learners formulate explanations from evidence" (p. 26). Explanations divorced from the evidence upon which they are based lose the persuasive power of scientific data, and teachers should let their students see as much real data as possible.

At this point, though, they shouldn't push students for a major change in beliefs. Teachers aren't running an evangelistic crusade, hoping to convert students after they attend only one service. Teachers should provide students with rich experiences with the evidence for evolution and give them time to process this new information. Student beliefs may not move a lot in the process of one unit, but biology teachers will plant a seed that studying about evolution is safe, interesting, and valid. That might be enough for students' first exposures. Even here in laying a foundation for future studies, however, biology teachers should not undermine students' faith. They should explicitly tell students that they are not trying to change their religious beliefs and back up what they say with true respect.

Model the difficult process of facing biases and conflicts of beliefs
Biology teachers who are themselves from religious traditions which conflict with evolution may become aware of ways they haven't processed conflict themselves. They should begin this journey themselves, especially if they want their students to do the same. Teachers won't understand what they're asking of their students if they haven't explored that territory personally. They'll also then be better able to support their students in this process.

Teachers who are not religious may also need to face their own biases. They need to consider if they have ever used evolution as a lever to pry a little bit of faith out of a student's grasp. Such teachers may have thought they were doing what was best for that student. They may line up with the skepticism of religion that Miller (1999) reports among some scientists by which "evolution leaves no room for God ... Our abilities to imagine the divine ... must exist [merely] because of natural selection. They surely do not exist because the Deity is real" (p. 179). Even teachers who haven't tried to divest religious students of their beliefs may still find value in examining their own beliefs. They should consider the value they place on their students' religions and how much they know about the lives of the religious students they teach. They should consider what they could learn about religious students and their families that would make them a better teacher of evolution as a topic and of students in general. Religion has been a powerful theme in human society, for good and for bad, and understanding it better will make for better teachers. Religion is even intrinsically intertwined in the history of the development of evolutionary theory itself (Good, 2003).

A note of caution is necessary here for secondary teachers: As adults, if teachers talk about their own views in front of the classroom, they may run the risk of subtle indoctrination. The process of making sense of conflicting religious and scientific beliefs is highly personal (Meadows, Doster, & Jackson, 2000). As Shipman *et al.* (2002) have found in studying how religious students make sense of stellar evolution, "Individual students have their own ways of approaching this topic ... Each of the [key students studied] has attempted to integrate an understanding of science and religion in their own way" (p. 543). High school biology teachers, then, will find a better role in empathizing with and supporting their students than in telling students the answers they've found.

Students need to struggle through the issues to find their own sense of management; hearing their teacher's view may short-circuit or unduly influence this process. A special note of caution needs to be given to literalist teachers who have nonliteralist students. Be careful of attempts to proselytize. Don't try to win them over to seeing the interplay of science and religion your way. This type of proselytizing is akin to a nonreligious teacher using evolution to divest students of their faith.

Role models can be a key support in the process of facing conflicts of beliefs. Teachers should consider connecting students with role models from the students' religions who can help them make sense of the tension. We know the power of role models in other areas of science. The first women scientists provided examples to future generations of female scientists. The more scientists of color we have in the profession, the more minority students are able to visualize themselves being successful in science. The same principle can be applied to religious students. Biology teachers can help their students network to find scientists from similar religious traditions. They can make sure that all of their students, not just the religious ones, know that many scientists are religious. One of my key mentors was both an evolutionary biologist and an elder in the church I was attending. Hearing him talk about the tension between his faith and his science, especially the tensions he still hadn't resolved, was the key for helping me move away from a focus on resolution and to one of management. He provided me an image of who I could be.

Consider teaching evolution as a case study in the nature of science

Evolution and its conflict with religious views of origins can be powerful avenues for helping students understand science as a way of knowing. My own sense of management began when I studied the nature of science in graduate school. I begin to see faith and science as different ways of knowing about the world. I remember thinking to myself, "Now, why for all these years did I think they even had to agree?" Religion and science approach the world with different assumptions and rules. Science looks for natural causes and accepts as evidence only natural events; faith looks for supernatural causes within the world of natural events.

Biology teachers may want to consider starting their evolution unit with a provocative statement: "We're going to learn about evolution, not Truth." They should help students separate scientific knowledge from any sense of absolute truth. Teachers must challenge any tendency within themselves to use the classroom as a platform for proclaiming truth in an absolute sense. Absolute truth is religion's area, not that of a science. Science knowledge is durable, but it is tentative (Rutherford & Ahlgren, 1990); nature keeps it from staking a claim as absolute. Some science knowledge is provisional, including some of the mechanisms of evolution.

Throughout the study of evolution, teachers should guide students to see how scientists make conclusions about the world through evidence and explanations. Teachers may even consider Good's (2003) strategy of helping students to see the inherent interplay that religion had in the development of evolution:

> Historical facts about the development of Darwin's theory of evolution by natural selection automatically include religion and religious beliefs … Helping students see the personal struggles Darwin had with religion, including his own beliefs, humanizes the development of his theory of evolution and places it within a real social context (p. 515).

As students grow in their understanding of the nature of science, teachers can help them to contrast the way humans have come to scientific and other forms of knowledge, including religion. This explicit instruction on the nature of science helps students understand the strengths and limitations of science. They also begin to see that human understanding is complex. It results from a rich interplay of multiple systems of knowing. Science is a powerful way of understanding the world, but it is not the only way of understanding the world. Focusing on science as a way of knowing can take much of the sting out of learning about evolution. Students have the freedom to opt out of learning when they feel threatened, realizing that science is not claiming to be absolute truth. When students raise objections to the conclusions biologists make about evolution, rather than arguing philosophically, teachers should guide students back along the trail of evidence from which those conclusions emerged. Teachers should guide students to see the logical flow without requiring students to internalize the conclusion. Teachers should allow students the grace of being illogical, especially since the evidence for evolution may create deep cognitive conflicts for religious students. When religious literalists see the reasoned flow of evidence and explanation that leads to evolution, they will probably be threatened, as I was. This threat may be of the caliber to shake their entire worldview, as Joanne's story describes. Teachers can ease the conflict some by reminding students that the discussion is not Truth in an absolute, religion-killing sense, but about scientific explanations based on evidence. This focus on the nature of science doesn't necessarily eliminate the conflict, but it can lower the stakes entailed.

Don't push, even if only toward management
Teachers of students from literalist backgrounds are probably aware of how high the stakes in this conflict are for many of these students. If some students take a strong stand for evolution, they will face serious repercussions, such as disapproval from their church or being ostracized by their families. In some religious circles, the lines between evolution and religion are drawn that staunchly. In light of this danger, teachers should not push their students into conflict management. Teachers should invite students into studying evolution, but not force or coerce students.

Teachers need to consider students' ages as they decide on their approach. Middle schoolers and ninth graders will probably find the study of evolution more threatening than eleventh or twelfth graders would. Delaying the study of evolution until the later years of high school may give students the emotional and intellectual maturity they need to tackle the topic.

Conclusion

For many students, studying evolution can be a threatening experience. Students who have grown up in a world similar to Joanne's typically see substantive conflict between their religious beliefs and scientific views of origins. Many of these students simply opt out of learning about evolution. They may memorize the information and represent it successfully on a test, but they haven't really engaged in the study of evolution.

Conflict resolution isn't an effective strategy with literalist students. Conflict resolution often forces these students into a dichotomous mentality where they feel forced to choose either science or faith. This forced choice may not even be their teacher's intention. Conflict management is the better approach because it allows students to reach their own comfort level in engaging in the study of evolution. This is especially true when biology teachers clearly communicate to these students respect and a desire to preserve students' religious beliefs throughout the study of evolution.

At its core, conflict management involves a human approach to the study of evolution. Conflict management recognizes that making sense of the world in the light of conflicting worldviews is at times difficult and messy. Management is difficult because two worldview systems, faith and science, give conflicting conclusions. It's messy because what works for one person often doesn't work for another. Conflict management recognizes that biology teachers can't and shouldn't solve this difficult problem for religious students, but they can be valuable mentors and guides, respecting students' religious beliefs and inviting them to take a further step or two into the study of the beauty and power of evolution.

References

Brickhouse, N. W., Dagher, Z., Letts, W. J. IV & Shipman, H. L. (2000). Diversity of students' views about evidence, theory, and the interface between science and religion in an astronomy course. *Journal of Research in Science Teaching, 37*, 340-362.

Good, R. (2003). Evolution and creationism: One long argument. *The American Biology Teacher, 65*, 512-516.

Gould, S. J. (1999). *Rocks of ages: Science and religion in the fullness of life*. New York: Ballantine.

Jackson, D. F., Doster, E. C., Meadows, L. & Wood, T. (1995). Hearts and minds in the science classroom: The education of a confirmed evolutionist. *Journal of Research in Science Teaching, 32*, p. 585-611.

Miller, K. (1999). *Finding Darwin's god: A scientist's search for common ground between God and evolution*. New York: HarperCollins.

Meadows, L., Doster, E. C. & Jackson, D. F. (2000). Managing the conflict between religion and evolution. *The American Biology Teacher, 62*, 102-107.

National Research Council. (2000). *Inquiry and the national science education standards: A Guide for Teaching and Learning*. Washington, DC: National Academy Press.

Plantinga, A. (2000). *Warranted Christian belief.* New York: Oxford University Press.
Rutherford, F. J. & Ahlgren, A. (1990). *Science for All Americans.* New York: Oxford University Press.
Shipman, H. L., Brickhouse, N. W., Dagher, Z. & Letts, W. J. IV (2002). Changes in student views of religion and science in a college astronomy course. *Science Education, 86*, 526-547.

The Personal and the Professional in the Teaching of Evolution

David F. Jackson

Introduction: Personal Cases, Professional Implications

> Facts are stubborn things; and whatever may be our wishes, our inclinations, or the dictates of our passion, they cannot alter the state of facts and evidence.
>
> John Adams

The scientific evidence supporting evolution is considered by virtually all scientists and by most science educators to be overwhelming in its variety and volume. Nevertheless, not only many science students but also many current and future science teachers personally disbelieve many of the "big ideas" that biologists and geologists consider central and nearly indispensable to their disciplines, including neo-Darwinian biological evolutionary theory, uniformitarian historical geology ("Deep Time") and Big Bang cosmology. This is often seen as intellectually unfathomable, professionally frustrating, and/or emotionally infuriating to many scientists and science educators. It is nevertheless true, at least in the US and particularly in the Southern region of the country, where it is no coincidence that the cultural and political influence of orthodox and evangelical Christian religion is strong, and perhaps growing stronger.

One obvious interpretation of the spirit of Adams' precept cited above might be the assertion and defense of a more rational and evidence-based scientific worldview (including acceptance of the historical fact of stellar, geological and biological evolution) against a caricature of more irrational and faith-based disbelief in those ideas. Its original context, however, was the American lawyer/patriot's principled but politically unpopular (and largely successful) defense of the British soldiers accused of the infamous "Boston Massacre." His effort was, at its heart, an exercise in looking beyond one's personal convictions and trying, in a professional capacity, to better understand and respect the coherence and complexity of the motivations and actions of those holding opposing views or different values.

Another stubborn fact is that, in discussions of evolution vs. creationism (or, more broadly, science vs. religion), people are often less than forthcoming about the personal histories, preconceptions, and perspectives that may significantly underlie their professional positions and practices. The phenomena of "hidden agendas" on the part of participants in the debate and "stealth" in their arguments are regrettably common. Therefore, in an effort to "do unto others as I would have them do unto me," I will provide a brief sketch of what I consider to be aspects of my own personal background that may provide a context relevant to a critical consideration of my professional comments.

Born and reared in the Northeastern US, I was exposed to ideas about evolution early, often, and in emotionally positive ways. Exactly like one of the paleontologists under whom I later studied and worked, "My father ... took me to see the Tyrannosaurus when I was five" (Gould, 1977, dedication), and I was nearly as fascinated by the abstract branching diagrams of evolutionary relationships on the museum wall as by the concrete and famously picturesque skeletons. Although unable to put the feeling into words at the time, as long as I can remember I have shared with Darwin the conviction that "There is a grandeur in this view of life" (1976, p. 484) in both an intellectual and (crucially, I think) an emotional sense.

While attending a very liberal, nominally Christian church and Sunday school with some regularity as a child, I absorbed from my parents an outwardly respectful but highly skeptical attitude toward religious scripture and doctrine. Much later I discovered that this basic point of view had been quite succinctly codified much earlier by Thomas Jefferson, who sought to separate highly valued ethical teachings from largely pernicious religious dogma and notions of the supernatural, as embodied in his personal, selectively expurgated version of the English Bible. At that time, I certainly already perceived that the worlds of science and of traditional religion were very foreign to each other. This point was driven home to me when I developed a strong intellectual conflict (regrettably, sometimes having personal and emotional overtones) with my closest teenage friend, who was raised in the conservative Jewish tradition but was inexorably turning increasingly orthodox. He took great delight in vehemently marshalling many of the same lines of argument against evolution used by creationists today, prominently including those now newly labelled Intelligent Design.

As a college geology major and research assistant in paleontology, a parallel skeptical streak led me to ask pointed questions that sometimes amused but often annoyed the several graduate students with whom I worked. None of them had ever seriously questioned issues such as, for instance, the logical basis of stratigraphic inference based on index fossils, the assumptions inherent in radiometric dating methods, or the plausibility of the development of vestigial organs driven by natural selection pressure for very subtle differences in energy requirements during growth and development. Although I was eventually fully satisfied of the soundness of scientific methods and conclusions in these and other areas, these concerns closely parallel some of more popular themes in current creationist rhetoric.

The personal views that I developed and currently maintain are essentially quite similar to those of unapologetic "scientistic" authors (e.g., Dawkins, 1998; Dennett, 1996)—placing great emphasis on the power of biological evolutionary processes, the vastness of geologic time and astronomical space, and the principles of the intelligibility of the natural world and of real progress in knowledge; with regard to religious belief, doggedly agnostic, bordering very closely on avowed atheism, viscerally contemptuous of all traditional dogma and of faith and/or revelation as ways of knowing.

My professional position on "the evolution/creationism controversy" is very simply what might be described as the "party line" among science educators: Darwinism, Deep Time, and the Big Bang should be taught, and various forms of creationism (all of whose origins, if we are to be honest, are of a religious nature) should not be taught, in US public school science classrooms.

So what subtleties are left to discuss? Several, from many different angles: political, personal, professional, and cultural influences, and some very real phenomena that, like many in the field of education, defy such categorization because they inherently involve complex interactions between these factors.

The Political: Problems Laid at My Doorstep

> *The devil went down to Georgia.*
> *He was lookin' for a soul to steal.*
> *He was in a bind 'cause he was way behind*
> *and he was willin' to make a deal*
> —Charlie Daniels

As a science educator at a public university in the American South, one reason that it is impossible to ignore the issue of evolution as a controversial aspect of science curriculum is the periodic bubbling up of the teaching of evolution (and/or creationism) as a political issue in my state and region. Although somewhat similar instances have occurred elsewhere in the US (e.g., the states of Kansas, Ohio, Pennsylvania, and, very recently, even the notoriously liberal New York), Georgia has been plagued with a recurrence of high-profile public controversies for the simple reason that this issue has great populist appeal and can be a highly effective way for a politician to achieve "name recognition," or general publicity of a largely positive nature, in the eyes of a great many voters. (All accounts in this section are based on articles appearing in the *Atlanta Journal-Constitution* and/or reported on Cable News Network and their web sites.)

In 1996, the state Superintendent of Schools, with an eye to what soon proved to be a campaign for Governor, made an appeal to what many Georgians saw as common sense when she formally and publicly wrote to the state's Attorney General, "If we teach only evolution, we are at cross purposes with what many parents teach at home and certainly what is taught in our churches. Do we have the right to do so?" On receiving the simple answer, "Yes," she proceeded to issue a follow-up statement that "education is a local issue, and we encourage teachers to decide for themselves how they are going to teach biology," clearly intending that people "read between the lines" her intent not to encourage the teaching of evolution by means of the (otherwise strongly emphasized) state curriculum standards. Many other prominent and aspiring politicians, including one ultimately elected as US Senator in 2004, immediately issued public statements in strong support of the Superintendent.

In 2002, the Board of Education in an affluent suburban Atlanta school district instituted a policy urging "discussion of disputed views of academic subjects, including human origins," and ultimately, in response to a threatened lawsuit by a creationist parent, required that the disclaimer in Figure 1 be placed in the front of every high school biology (and middle school life science) textbook:

> This textbook contains material on evolution. Evolution is a theory, not a fact, regarding the origin of living things. This material should be approached with an open mind, studied carefully, and critically considered.
>
> *Approved by*
> *Cobb County Board of Education*
> *Thursday, March 28, 2002*

Figure 1. Textbook disclaimer sticker mandated by the Cobb County (Georgia) school district.

Despite protests from many teachers in the district, ridicule from scientists and science educators at the state's most prominent universities, and some politicians' publicly stated fears that the state's long-term economic interests would be harmed by damage to the state's perception in national and international public opinion, this policy was perceived to be at least tacitly supported by a majority of the district's population. In response to a countersuit by another parent, the policy was ultimately found to be unconstitutional by a federal appellate court in January 2005, but it was understood that the disclaimer stickers were not to be removed until at least the end of the school year.

In 2004, a different State Superintendent of Schools proposed striking the word "evolution" from the state's newly revised science curriculum standards and replacing it with the phrase "biological changes over time." She said the concept of evolution would still be taught under the proposal, but the word would not be used. She repeatedly referred to evolution as "a buzzword that causes a lot of negative reaction" because many people associate it with "that monkeys-to-man sort of thing," and said that her purpose was to alleviate pressure on already overburdened teachers in socially conservative areas where parents object to its teaching.

As in 1996, the proposal drew immediate harsh criticism from many, this time including former Georgia Governor and U.S. President Jimmy Carter, whose deeply held evangelical Christian faith has been a well-known fact of American political culture for 30 years:

> As a Christian, a trained engineer and scientist ... I am embarrassed by [the Superintendent's] attempt to censor and distort the education of Georgia's students. Nationwide ridicule of Georgia's public education system will be inevitable if this proposal is adopted ... The existing and long-standing use of the word 'evolution' in our state's textbooks has not adversely affected Georgians' belief in the omnipotence of God as creator of the universe ... There can be no incompatibility between Christian faith and proven facts concerning geology, biology, and astronomy ... There is no need to teach that stars can fall out of the sky and land on a flat Earth in order to defend our religious faith. (CNN, 2004)

While Carter is a personally beloved and respected figure among many in the state and region, his view of the unproblematic independence and compatibility of science and religion (cf. Gould, 1999) hardly represents the mainstream of either public opinion or political rhetoric.

The Superintendent's statement also drew an immediate public response from the current, newly elected Governor, which was portrayed in the media as a sharp reprimand that elicited a public apology. In response to the Governor's statement that "As public officials, we don't have the luxury of thinking out loud; I believe that's what she was doing," the Superintendent responded that "I misjudged the situation and I want to apologize for that." This was backpedaling only in a political sense, however, not a substantive one. The Governor continued, "If you're going to teach evolution, you ought to call it evolution. By that I mean, there ought to be a balance. Evolution, as I understand it, is an academic theory. I think it should be taught as academic theory." In ensuing interviews in which the Superintendent was asked to clarify her reasoning, it became apparent that the former social studies teacher's own sincere understanding of "evolution" was that the scientific term applies only to the progressive history of "lower" forms of life, and that naturalistic notions of either the origin of life or of human ancestry among earlier anthropoid species are, scientifically speaking, misconceptions. Although the proposed standards are fairly strong and explicit in many ways, they do strategically omit any reference to these most controversial implications of a fully naturalistic, scientific view of evolution for the most remote and most recent events.

The "take-home message" of the incident, in the minds of the state's voters and teachers, was that both the Superintendent and the Governor personally disbelieve evolution and implicitly support initiatives like the disclaimer stickers and, if it were proposed, an "equal time" or "balanced treatment" law of the type legislated and declared unconstitutional in the past in other Southern states. Soon after, in the broader context of other controversies involving "faith-based" public policy decisions, the state's more senior U.S. Senator placed in the Congressional record his belief that "there is no separation of church and state in the Constitution."

While it is relatively easy for scientists and science educators to glibly dismiss such political posturing as nonsensical or beneath contempt, the publicity garnered by these statements and the hint of legal and intellectual legitimacy that they lend to creationist political programs has a real effect on the minds of my students.

The Personal: "I still just don't buy it."

> The truth of a proposition has nothing to do with its credibility. And vice versa.
> — Robert A. Heinlein

Another issue that confronts me daily is that many (perhaps even a majority) of my students (prospective middle school science teachers, predominantly in their junior year of college) personally disbelieve evolution, often but not always on the familiar grounds that they perceive it to be in conflict with their (nearly uniformly Christian) religious backgrounds. Although obviously a somewhat academically select and professionally self-selected group, nevertheless in this way they roughly reflect the views of US adults as a whole.

> Extreme prejudice against evolution based on religious precept may be common, but is certainly not the only explanation for widespread disbelief. A widely publicized and annually repeated Gallup poll of US adults bears this out. It has been known (and poll results have been remarkably consistent) for over 20 years that about 45% of Americans believe that "God created human beings pretty much in their present form at one time within the last 10,000 years or so." Extra questions included in the survey for the past three years, however, show that less than half of these people (20% overall) also believe that "the Bible is the actual word of God and is to be taken literally, word for word." The demographic group in which disbelief in evolution not paired with Biblical literalism is most common is 18—29-year olds. (Newport, 2004)

Certainly, there are many people who think that they cannot even remotely consider the possibility that the basic "story" of evolution is true because doing so will lead to their eternal damnation. There are also many, however, who have given at least some consideration to the issue, and whose initially milder misgivings are strongly reinforced by the literature of "Creation Science" or "Intelligent Design," now more readily available than ever, of course, via the Internet. This literature often is more convincing to "inquiring minds" than are science textbooks or teacher lectures, because it makes more explicit, immediate and strategic links between its claims and the alleged evidence. Despite some shining counterexamples and undoubted overall improvement in recent decades, science textbook presentations (and therefore the didactic efforts of many teachers) typically do not centrally focus on evidence-based reasoning and do a relatively poor job of "building a case" for the basis in tangible evidence for the litany of facts and theories with which they are overstuffed. Highly focused and coherent refutations of creationist literature exist and continue to be produced, and some of these (e.g., Eldredge, 2001) are written in a distinctly less strident and somewhat more conciliatory and hopeful tone than comparable earlier semi-popular efforts (e.g., Eldredge, 1982), but nevertheless are less accessible and less convincing to readers whose preconceptions are strongly opposed to evolution.

Such questioning of the "finished science" conclusions presented in textbooks might be viewed as a commendable instance of highly appropriate skepticism, one of the "scientific habits of mind" touted by the various national standards documents, if it were truly motivated by a desire to critically understand the evidence. This is not often the case, of course, as witnessed by the well-documented selective consideration and distorted emphasis on anomalous data commonly used in "Creation Science" literature. In combination with my own experiences recounted earlier, however, it does make me more inclined than I otherwise might be to attribute at least partially nonreligious motives to at least some of my students' pointed questions about transitional forms, the fragmentary nature of the fossil record, etc.

Among my own students, a common stumbling block is that their own academic exposure to evolution, typically restricted to the context of a biology course unit centered on the basic mechanisms of genetics, consists nearly entirely of classic modern examples of microevolutionary phenomena of adaptation within species, or possibly genera, many involving artificial selection or unintended human influence: peppered moths; antibiotic-resistant bacteria; breeds of dogs; perhaps Darwin's finches or tortoises. A common and highly understandable conclusion is that microevolutionary "fine-tuning" is very plausible (indeed, sometimes even clearly observable over a relatively short time period), but that this kind of evidence does not basically contradict the intuitive appeal of the venerable Neo-Platonic notion of "natural kinds" or of teleological views of humankind as not only the current apotheosis of existence or pinnacle of evolution but also as its preordained or actively directed goal. The nagging feeling remains that the leap of extrapolation to macroevolution, geologic Deep Time, fully naturalistic explanations of natural phenomena, and the necessary relationship between them is in the nature of abstract and abstruse scientific "hand-waving." In this situation many students will "compartmentalize" their academic knowledge, "learning it for the test" or even "telling the teacher what s/he wants to hear," without taking the concepts seriously as true or as relevant in any way to the "Big Picture" of their everyday lives.

Surprisingly to many teachers, the American Association for the Advancement of Science's excellent and highly thoughtful *Benchmarks for Science Literacy* (1993) argues that full-blown Darwinian theory about evolutionary mechanisms should probably not be taught in middle school. Their reasoning is not that it is potentially controversial or not sufficiently important, but that it is too abstract, and perhaps implausible, for students of this age range. In this view, students would be better served if middle school science emphasized the large, varied, and often individually fascinating body of biological and geological observations, as they historically accrued, and how scientists came to view this evidence as first suggesting, and then increasingly strongly supporting, the general idea of evolution. In other words, according to Gould's (1983) oft-misinterpreted distinction, the evidence for the basic "fact" of evolution (the scientific version of the "story" of major events in the history of the Earth and of life) can and should be considered separately from, and earlier than, the detailed "theory" of evolution (neo-Darwinism, ultimately including details of genetics and basic biochemistry). AAAS's recommendations regarding quantification of absolute geological time and Big Bang cosmology are in a very similar vein—the basis for these scientific conclusions can be expected to become fully intelligible only in high school, after sufficient prerequisite background knowledge (notably a deep understanding of atomic structure) has been developed first.

Another very real problem is that the nature of scientific knowledge and investigation in historical geology, macroevolutionary theory, and cosmology is very different from "textbook" scientific method, with its heavy emphasis on controlled experimentation, replicability, and prediction in a narrow sense of predicting future events. As a result, even the relatively few students who take their typically limited instruction in the nature of science seriously find the history of the universe, of the Earth, and of life less comfortable, reliable, and credible than other scientific conclusions. Specific conclusions about evolution are not only tentative (like, strictly speaking, all scientific knowledge), but based on unfamiliarly historical, inferential, and in some ways inherently ambiguous (Ault, 1998) patterns of reasoning about unfamiliar kinds of evidence.

With my students, however, sometimes the problem is infinitely less subtle than any of these. Due to the quantity, quality, or both of their own science education, many are very simply ignorant of some basic facts and basic procedures of science, most often in the field of earth science, and often particularly having to do with fossils and historical geology. I discovered this one year when my students were assigned, in their Reading course, to identify and evaluate trade books in each major subject designed for very low-level readers among our target age of grades 4 through 8. One of them came to me in frank amazement at how much they had learned from a pair of books about fossils intended for a first- or second-grade reading level (Aliki, 1988, 1990). They had just completed a unit on the same topic in the Geology course which is a co-requisite of mine, in which they had memorized many highly detailed facts about particular fossils, but managed not to understand any general principles about their formation and use whatsoever.

The Professional: A Snake Coiled Up under the Table

> In the beginning the Universe was created. This has made a lot of people very angry and been widely regarded as a bad move.
>
> — Douglas Adams

At least in the US, the conflict with which we are concerned here will eventually rear its head in any broad and deep discussion of science, education, or religion, alone or in combination, as well as in politics and in the minds of individuals. Both religion and science (introduced, by the way, in that historical order) have been so central to the collective national experience and culture of the US that their differences have long lurked beneath the surface of discussions of seemingly unrelated issues. Historians have often said the same about the pervasive influence of racism as a force in American constitutional and social history, a conflict that, not entirely coincidentally (Menand, 2001), also most prominently came to a head in the years immediately following the publication Darwin's *Origin of Species*, but likewise continues to the present despite the perception that "the basic issue was settled" some time ago.

Middle school teachers cannot realistically avoid the potential for the perceived conflict by ignoring evolution. The spotlight in this conflict tends to be on high school biology, where "The E Word" is usually explicitly taught as a distinct unit (or, much more rarely in practice, as a *leitmotif* of an entire course). However, evolution, broadly defined, is central to earth science as well and is intimately related to other "uncontroversial" topics in basic life science. Every earth science curriculum and text will cover the existence and basic types of fossils, the gradual and cyclical nature of geologic processes, the outline of the geologic timescale, and the diversity and inferred life cycle of stars. Life science always includes adaptations of organisms to their environment, basic phenomena of heredity, and biological classification as reflecting nested patterns of morphological, behavioral, and ecological similarity as well as diversity. At least some students will (and should) make the connection between these categories of facts that are always part of a middle school curriculum and the Big Idea behind them, which is evolution. Indeed, in my region, some will have been actively trained to do so and coached, by their parents or religious leaders, as to what critical questions to ask or protests to register. Ideally, of course, making such connections should be a goal of teaching at this level, and it is one too often neglected. It also means, however, that teachers who hope to let evolution "slip by" in the name of avoiding controversy are very unlikely to be able to do so if they are making even a partial attempt to cover their curriculum.

Often the specific terms "evolution," "Darwin," "natural selection," "Big Bang," etc., do not actually appear in middle school curriculum guidelines (although in many states they do) or, often, in textbooks. Euphemisms like "family relationships," "historical development of organisms," "change over time," "theories of origins," or (in the state of Georgia's own former peculiar phrasing) "the ability of organisms to change as necessity [*sic*] for species survival," are commonly substituted, but the intention is usually clearly that some notion of scientific views of evolution be taught.

Is a Conflict Really Inevitable?

> Both may be, and one must be, wrong. God cannot be for and against the same thing at the same time ... it is quite possible that God's purpose is something different from the purpose of either party.
>
> — Abraham Lincoln

Lincoln, of course, was speaking about slavery and the American Civil War, and perhaps we risk clouding the evolution controversy by comparison, but I believe that the parallel is instructive again. A key to making any progress at all in understanding differing points of view is to admit *in spirit as well as in principle* that not only may none of us have a monopoly on truth, but that some sort of middle ground position may be tenable. To people at both extremes of this debate, not only does controversy seem inevitable, but it is also self-evident that there is an unresolvable intellectual (and perhaps also emotional) conflict between evolution and creationism, or more broadly between science and religion as ways of knowing. There is, however, a broad array of points of view about the relationship between science and religion, and in a crude but heuristically useful way, many can even be viewed as part of a linear continuum of shades of belief (Scott, 2000). Some scientist–commentators (e.g., Gould, 1999) maintain that both the ways of thinking and their appropriate purposes are so separate that in principle there can be no conflict because there can be no meaningful interaction. Others (e.g., Miller, 1999; Skehan & Nelson, 2000) argue that evolution and quite traditional (although not scriptural literalist) religious beliefs can truly coexist and be reconciled by individuals with a sincere concern for intellectual honesty. Others (e.g., Barbour, 2000; Haught, 2000; Polkinghorne, 1998) even argue that science and religion, albeit with one or both conceived of in a very specific, subtly circumscribed way, can actually constructively and synergistically interact with each other.

I have found that most of my students are unaware of the fact that most major "mainstream" church hierarchies (e.g., Episcopal, Methodist, Presbyterian, Roman Catholic) have issued formal statements acknowledging that voluminous and highly convincing evidence for evolution exists, and stating as a general principle that there is no necessary conflict between evolution and Christian faith. Many, however, are aware of this but do not care—Americans are notorious for coming to independent conclusions about the practical implications of their faith regardless of the official statements of the nominal leaders of their denomination. Indeed, it is increasingly apparent that, in the US, differences between more conservative and more liberal (or traditionalist versus progressive) "camps" or "wings" of a given Christian denomination are of much greater practical significance than are the often subtle official doctrinal differences between denominations (Rosin, 2005). It is also clear that close scrutiny (e.g., Dawkins, 2003) of some of the carefully crafted statements of church leaders, most notably that of Pope John Paul II (1996), often reveal only a nominal and superficial acceptance of evolution, accompanied by strong and unambiguous qualifying statements explicitly positing the existence and crucial importance of supernatural phenomena or specifically supporting the notion of active and conscious divine intervention at either or both of two psychologically crucial points in Earth history: the origin of life and the first appearance of humans ("the moment of the transition to the spiritual").

Although my own students are nearly all Christian, the same phenomenon is apparently common among Jews and at least some Muslims. Clear and strongly worded statements authored by the American Jewish Congress and the Central Conference of American Rabbis vehemently denounce the political efforts of creationists and support the teaching of evolution in public schools, yet many American Jews are also strongly personally committed to a literal interpretation of Genesis. Evolution is considered anathema in many Muslim nations, and American Muslims are often thought of in the same light as Christian or orthodox Jewish "fundamentalists" on this issue. Perhaps ironically, at least one form of Islam, the Twelver Shia school of thought (the dominant sect in notoriously theocratic Iran), actually takes a middle-ground position on evolution and creation (Ahlul Bayt Digital Islamic Library Project, 2005) that I find strikingly similar, in substance, to that of the Pope.

Whose Life-and-Death Issue?

> Creationists tenaciously cling to the wisdom and worldview of a Near Eastern culture thousands of years old ... So creationism seems to me to threaten the integrity of our children's education, and thus threaten the long-term well being of our country.
> — Niles Eldredge

> Christianity, if false, is of no importance; and if true, of infinite importance. The only thing it cannot be is moderately important.
> —C. S. Lewis

At a personal level, scientists and science educators tend to believe that science in general, and an understanding and acceptance of evolution, Deep Time, and/or the Big Bang, in particular, is an absolutely vital part of being an educated person, or even an intelligent or basically rational person. While I personally understand the great appeal of this mode of thought, I would argue that in the context of the larger debate, what I and some of my closest colleagues like to call "The Big Conversation," scientists and science educators must realize that religion is a highly positive, constructive, and adaptive force in many people's lives at a practical level, and should be taken seriously for that reason alone. In the minds of many people of perfectly sound basic intelligence and good will, if they are told that they must choose between their faith and science, there is no contest (and religion "trumps" science). While an excellent argument can be made that the history of ideas in Western civilization largely represents the gradual and inexorable march of the progress of science and the corresponding eclipse of orthodox religion, the appeal of the prospect of certainty and simplicity, and its utility for the daily lives of many people, remains strong.

It is easy to forget this basic, stubborn fact about even the "Modern" world, and those with a thoroughly secular outlook on life, inclined to put reason and science on an intellectual pedestal (a category in which I include myself), do so at the peril of grossly misunderstanding, and being considered irrelevant by, a large part of the US population (Brooks, 2003). In framing the general argument in the context of teaching and teacher education, I emphasize the serious consideration of "middle ground" positions espoused by some respected scientists (e.g., Miller, 1999) not because I personally find their arguments convincing (which I do not), but because I believe that they are, in practice, the only hope of serving as a "wedge" to initiate at least a minimal opening of a substantive and potentially productive interaction between people with polarized views.

Other People's Children

> If I saw a venomous snake crawling in the road, any man would say I may seize the nearest stick and kill it. But if I found that snake in bed with my children that would be another question. I might hurt the children more than the snake, and it might bite them. Much more, if I found it in bed with my neighbor's children, and I had bound myself by a solemn oath not to meddle with his children ... it would become me to let that particular mode of getting rid of the gentleman alone.
>
> — Abraham Lincoln

The canonical literature of multicultural education focuses primarily on issues resulting from prominent differences between people in characteristics like race/ethnicity, language, gender, or socioeconomic standing. As I have argued elsewhere (Jackson et al., 1995), however, differences in beliefs and experiential backgrounds between superficially similar people can sometimes be so pronounced as to constitute a *bona fide* cultural gap that needs to be painstakingly bridged. An example of such a group of people is my own undergraduates, almost uniformly middle-class, academically successful, native-English-speaking, white female US citizens, with a strong interest in science.

Applying the spirit of Delpit's (1995) practically rooted recommendations for responsible teaching of people (especially children) of a distinctly different culture, we need to consider at least the possibility that the substance or style of education that is or was appropriate for people like ourselves may not be universally applicable. Again taking a cue from Lincoln in the much more weighty context of 140 years ago, belief in our own righteousness should not necessarily translate into a "Holy War" on behalf of science that might inflict significant "collateral damage" in the name of a worthwhile cause.

What might be the professional implications of a personal viewpoint dominated by secularism and a strong emphasis on science? It is explicitly or implicitly axiomatic to many scientists and science educators that a creationist of any stripe (and there are most certainly many versions of creationism) cannot be a good science teacher and should not become one. My closest and most highly respected colleague, who is also very much a personal friend, strongly believes this. We can make excellent arguments for the overwhelming nature of the scientific evidence for evolutionary thought, and thus argue that it is an important set of understandings that should be included in any reasonable curriculum. But can we make this value judgment about students' beliefs, as opposed to their understanding (Cobern, 1994; Smith, 1994)? I would maintain that, in the modern multicultural world of education, we cannot.

To me, this principle is even more applicable in the context of science and teacher education at the middle school level. I have consistently found that current and prospective middle school teachers, compared to their high-school-level peers, are weaker in science content background, less inclined to consider themselves as science teaching specialists or to see science as a central or crucial aspect of school curriculum, and less likely to exhibit what the AAAS *Benchmarks* calls "scientific habits of mind" and to consider that mode of thought as an inherently positive aspect of their personal life. While middle school science teacher education can and should aim somewhat to alleviate some of these deficiencies, we must bear in mind that the role of science in general and evolution in particular in a middle school curriculum is widely perceived to be different from that role in high school science. In the minds of most middle school teachers, principals, and the most influential curriculum theorists (e.g., Beane, 1993), and in opposition to the explicit or implicit views of perhaps most scientists and science educators, science is only a very small and (at best) coequal part of an ethos of student-centered curriculum and teaching, and not so much a privileged separate subject or a way of knowing that is seen as having unique value. Its status as a distinct part of the curriculum is more an artifact of teacher staffing issues (the severe shortage of highly qualified science specialists among middle grades teachers) than of principled curriculum design. If "pushing" the crucial importance of evolution is judged likely to result in student alienation from a teacher (cf. Kohl, 1992) or a broader subject area (science in general!), it may more justly be considered expendable, or a least a low priority, than in the case of high school biology. To many excellent and dedicated middle school science teachers, "I don't teach science, I teach kids" is not just a trite slogan (or perhaps, to put it in a slightly more positive light, a somewhat oversimplified abstract statement about constructivism) but is truly a way of professional life.

Whose Responsibility, to Whom?

> I do perceive here a divided duty.
> — William Shakespeare

> The greatest conflicts are not between two people but between one person and himself.
> — Garth Brooks

As a general guiding principle, can science teachers follow Polonius's (*Hamlet*) old saw, "This above all, to thine own self be true"? Should teachers with creationist views use their own classroom setting to "bear witness" to their own beliefs, including those of a religious nature? Surely not, nearly every scientist and science educator would say. But the principle cuts both ways. Many scientists and science educators may resonate deeply with some of the more eloquent and vitriolic criticisms (e.g., Dawkins, 1998 and elsewhere) of the grossly unscientific (and especially religious) worldviews held by many people, and might consider this attitude to be an inherent aspect of science itself. They may sometimes, in weak moments, even express these views in a self-righteous tone worthy of any television evangelist, in "preaching to the choir"-type conversations with known like-minded people. Very occasionally, I do so myself. But not in front of my students! Some teachers can, and often should, partly or even largely suppress their personal beliefs if they judge that they are not in keeping with responsible professional practice. At the very least, they should most certainly keep their immediate, visceral private feelings from influencing their carefully considered public expression.

I know several excellent middle school earth science teachers in Georgia who do not allow themselves to act on these impulses with regard to their own creationist beliefs, teaching in a very straightforward way about evolution, uniformitarianism, Big Bang cosmology, etc., because it is part of the curriculum in their state and district. Most feel a very real internal conflict about this. Some feel guilty about not "spreading the good news" of their religion, which may include strict, Biblical literalist creationism, but accept the responsibility not to do so.

For several years, I had a former student of this description, a dynamic and innovative earth science teacher who returned to my class as an invited guest instructor, openly telling her personal story and beliefs, explaining her sense of professional duty and her classroom practices, and answering questions from many students with similar personal beliefs. On several occasions I held her up as a shining example in conference presentations and discussions with colleagues, especially scientists. I was appropriately rewarded for my hubris in this matter when she decided recently that she could no longer reconcile her compromise with her conscience, resigned her position as a public school teacher, and moved to a conservative Christian school in which the Bible is explicitly used as a science text on this subject.

Likewise it is not my role to "convert" my education students, but to lay before them the difficult choices that they must make for themselves. For creationist and/or relatively scientifically ignorant students, this certainly means trying to teach them in the most effective possible way about evolution and the strong evidence on which it is based. But for students (or teachers, or teacher educators) who have always had a strongly scientific cast of mind, this means raising their consciousness about the very real reasons that "those other people" exist and probably will continue to do so, both among their students and among their colleagues.

In this I have also tried to enlist the help of scientists on my campus whom I know to also be practicing Christians, hoping that as invited guest instructors they might function as role models of a sort that I cannot personally provide. The results have been mixed. On the one hand, the scientist who bitterly proclaimed, "Biblical literalism is not just bad *science*, it's bad *theology*" did not contribute to the kind of atmosphere I try to create in my classroom. On the other, the one who presented a highly detailed overview of the anthropoid fossil record and the ongoing revision of theories about the human family tree, while prominently wearing an oversized wooden cross pendant that the students could not possibly fail to notice, was judged to be one of the highlights of the course.

In recognition of the *de facto* major role of religion in American society, Nord (1995) proposed that religion and/or religious ways of knowing, somehow taught in a manner that scrupulously does not favor one faith or sect over any other ("Ah, there's the rub!"), should ideally be infused prominently throughout all subjects in the curriculum in American public schools. At the very least, he argues for one or more required core courses on religion, alongside science and other academic subjects, without being confounded with them.

Recently a science educator (Anderson, 2004) has reached similar conclusions from a very different theoretical basis, Postmodernism in general and the conception of Curriculum as *Currere* in particular. He endorses Nord's basic suggestions ("Religion must be taught objectively and neutrally," p. 158) and specifically applies them to the case of evolution and creationism. Like nearly all science educators, he supports boldly presenting evolution (and not any form of creationism) as science, but he also pictures pre-college science teachers subtly and creatively helping young students to "not only extend the interface between disciplines but concern [themselves] with synthesizing core subject matter with personal understandings, including religious ones, in autobiographical fashion." (p. 84) I find this vision of ideal teaching "if fear were replaced with courage" (p. 91) usefully theoretically challenging, and it squares very well in practice with what I try, tentatively and perhaps feebly, to do in my own college-level Science Education course (described below), but I remain unconvinced that it is remotely realistic in the applied context of US public school science classrooms.

W(hat) W(ould) J(ohn) D(ewey) D(o)?
What Seems to Work for Me in My Classroom

> *Really don't mind if you sit this one out.*
> *My words but a whisper, your deafness a shout.*
> — Ian Anderson

How might a science teacher educator, committed in general outline to the spirit of the progressive principles of Dewey (e.g., 1916, 1933, 1959), relate to science education students sporting "W(hat) W(ould) J(esus) D(o)?" bracelets, as many of mine have for years? Good (2005) argues that Dewey's stress on the truly central role of science in a school curriculum in a democratic society makes this controversy a zero-sum game—strong support for the development of scientific habits of mind implies an explicit stance in opposition to religious habits of mind, or at least an effort to explain away the psychological phenomenon of religious belief in naturalistic, scientific terms.

Some of my scientist colleagues here in Georgia make some attempt to engage their creationist students in constructive dialogue. Most, however, consistently come to the conclusion that, since no kind of meaningful compromise on either side is really possible (e.g., Palevitz, 2003), people with different views tend merely to "talk past one another," and there is no hope of a practical solution to the personal intellectual and emotional conflict. To many (e.g., Palevitz, personal communication), the appropriate professional response to this on the part of scientists and science educators, in the downcast spirit of Anderson's lyric, is simply to discourage such students from further pursuing science or science teaching.

My experience with my own students over the past 17 years has led me to a somewhat different approach to "confronting the elephant directly" (Good, 2005, p. 48). I will now outline several examples of the way I teach about evolution and "The Creation Controversy" (Skehan & Nelson, 2000) in my own science methods course for prospective middle school teachers, in an effort to subtly nudge them towards engaging in experiences that give them the opportunity to better understand evolution as an important exemplar of science in general. Details of the scope and sequence of this course, and specific class handouts used, may be found at: http://djackson.myweb.uga.edu/ESCI4430.html.

First, although I am supposed to be teaching about teaching, I do have the opportunity to teach some science in the course of doing so. In introducing the topic of teaching "basic science process skills" (observing, describing, inferring, communicating, etc.), we do a laboratory activity (based on Lawson, 1995, pp. 186–188) on comparative anatomy of animal skulls and how it relates to their adaptations to survival. In introducing the possible role of video-based curriculum packages in teaching, we view a segment of *The Voyage of the Mimi* (starring the middle-school-aged Ben Affleck) focusing on paleontological evidence for the evolution of whales. In the course of a hands-on introduction to the practical advantages and disadvantages of dissection activities, we note the homology of the bones and muscles in chicken wings to those in our human arms. To introduce the strategy of scale modeling, we use a 40m cash register tape to represent the last 4 billion years of the geologic timescale. My students are also asked to consider and discuss some carefully selected videotape excerpts (from Burke, 1986) about the history of science and its relation to religion, and to read the classic summary of the evolution of ideas about fossils and geologic history in Prothero & Dott (2003) or (preferably, in my opinion) its earlier, more detailed incarnation in the now-out-of-print Dott & Batten (1988). (An excellent alternative source for much of this historical information is http://evolution.berkeley.edu/evosite/history/index.shtml.)

Second, discussion of evolution, religion, and the relationship between them may be better accepted by students when presented in a larger context of ethical decision-making in one's personal and professional life, and only after the students and teacher have established a mutual trust and comfort level in open discussions. If evolution were the only topic around which I introduced these issues of personal versus professional views, some students might feel persecuted by the specific focus. In my course, this issue is the last among many in this general theme, with earlier discussions of sex education curriculum policies, the ethics of the use of living and dead animals in the lab, and environmental issues having paved the way for an atmosphere in which conflicting ideas can be laid out and discussed in a mutually respectful and constructive manner. Responsible professional consideration of these other topics may also involve the above-mentioned active suppression of personal beliefs, whether of the left or of the right.

Third, our specific and explicit consideration of evolution and creationism issues follows a careful sequence of several phases:

- If applicable, as it has been the past few years at this writing, begin with news reports (print and/or video) about the latest political battle over this issue in the state or nation, as a way of generating interest.

- Look at the latest available poll results about the beliefs of US adults about evolution and creationism (e.g., Newport, 2004, as cited earlier).

- Survey relevant legal precedents (Matsumura, 2001), the substance of which is surprising to many of the students because of their own experiences as high school students, most often in Georgia. Usually, several have been exposed to creationist tracts in their science classes, many have had teachers who explicitly shared their personal beliefs with the students (most creationist and fewer emphatically anticreationist), and many simply were never taught evolution at all, with some assuming that it was illegal to do so.

- Specifically acknowledge and address some of the more common (and, to some degree, understandable) misconceptions about evolution and about stereotypical linkages between beliefs on this issue and other aspects of a person's character (cf. Brem, Ranney, & Schindel, 2003).

- Introduce Scott's (2000) continuum of beliefs diagram ("Flat-Earther" through "Materialist Evolutionist"). It is at this point that, for the first time, I explicitly invite students, if willing, to share their own beliefs. Invariably, we find that many of them would fall very much in the middle region of the continuum.

- The culminating activity of the unit is a discussion of the relative merits of the most common practices followed or advocated as solutions to The Problem (Figure 2). I do not use, and actively discourage the use of, the term "debate" as connoting a necessarily adversarial relationship.

Teaching Evolution: Beginning of Final Discussion

Below are many very different responses to the problem of teaching about evolution in a context in which some students feel strongly that it conflicts with their religious beliefs. I have known multiple teachers, at various grade levels, who have actually used each of these approaches. Do any of them appeal to you as possible solutions? Why, or why not?

- Assume that this problem is not important, or pretend that you are not aware of any potential problems of this nature.
- Don't explicitly teach evolution, etc., at all, or greatly de-emphasize it in comparison to its importance in curriculum guidelines.
- Make a brief introductory statement or "disclaimer" at the beginning of these studies, and then proceed as normal.
- Excuse students who are "conscientious objectors" from class for these days or units.
- Allow such students, on their own initiative, to state their objections or contrary beliefs in class, and allow or encourage some degree of discussion of the problem in this situation.
- Give "equal time" (and thus equal status) to teacher presentation of evolution and of various forms of creationist views in science class.
- Encourage an open discussion of the issues, including sharing your personal views.
- Be open to discussing the issue with students, but only outside of regular class time.
- Use a presentation of creationist views specifically as a prime example of what is *not* science, with the stated objective being to teach about the nature of scientific knowledge.
- Include a presentation and/or discussion of scientific vs. religious views and "ways of knowing" as part of an explicitly interdisciplinary unit (science and social studies, for example), therefore getting around the "strictly science in science class" principle.

Figure 2. Excerpt from class handout:
Proposed practical solutions to the problem

In a typical year, at least one student strongly supports every one of these possibilities. Often their stated rationale is related to the approach of a teacher they have known in the past, viewed either as a model or as an example of a crucial mistake to avoid. The last option tends to be particularly popular, perhaps because the Middle School Concept (Beane, 1993) is so strongly emphasized in the core courses in their teacher education program, and that concept is often grossly oversimplified as "anything interdisciplinary is inherently good, other things being equal." I have yet to see or hear of it actually being carried out by multiple teachers on a middle school team.

Postscript: Is There Any Hope of Progress in Understanding Each Other?

Will a dialogue among people of goodwill, and a greater understanding of the viewpoints and values of others, eventually lead to greater personal self-understanding, carefully considered professional responsibility, and political peace? Consider the following passage:

In the gospels, Jesus went out of his way to explain to his followers that his Kingdom was not of this world, but could only be found within the believer ... In the modern West, we have made a point of separating religion from politics; this secularization was originally seen by the *philosophes* of the Enlightenment as a means of liberating religion from the corruption of state affairs, and allowing it to become more truly itself. But ... religious people ... often feel that they have a duty to bring their ideals to bear upon society... If state institutions did not measure up to the [scriptural] ideal, ... or if their community was humiliated by apparently irreligious enemies, [a faithful person] could feel that his or her faith in life's ultimate purpose and value was in jeopardy.

(Armstrong, 2000, pp. x–xii)

How would a thoroughly secular scientist react to this? How would an evangelical Christian in today's US react? Would it make a difference to either person if they were told that the religion and the region that the author is considering is not Western Christianity but rather Islam in the context of the current political situation in the Middle East?

References

Aliki (1988). *Digging up dinosaurs (And putting them together again)*. New York: HarperCollins.
Aliki (1990). *Fossils tell of long ago*. New York: HarperCollins.
Ahlul Bayt Digital Islamic Library Project (2005). *Man and evolution*. Available at http://al-islam.org/philosophyofislam/11.htm.
American Association for the Advancement of Science (AAAS). (1993). *Benchmarks for science literacy*. New York: Oxford University Press. Available at http://www.project2061.org/publications/bsl/online/bolintro.htm.
Anderson, R. D. (2004). *Religion and spirituality in the public school curriculum*. New York: Peter Lang.
Armstrong, K. (2000). *Islam: A short history*. New York: Modern Library.
Ault, C. R., Jr. (1998). Criteria of excellence for geological inquiry: The necessity of ambiguity. *Journal of Research in Science Teaching*, 35, 189–212.
Barbour, I. G. (2000). *When science meets religion*. San Francisco: HarperCollins.
Beane, J. A. (1993). *A middle school curriculum: From rhetoric to reality* (2nd ed.). Columbus, OH: National Middle School Association.
Brem, S. K., Ranney, M., & Schindel, J. (2003). Perceived consequences of evolution: College students perceive negative personal and social impact in evolutionary theory. *Science Education*, 87, 181–206.
Brooks, D. (2003, March). Kicking the secularist habit. *The Atlantic Monthly*, 291(2), 26–28.
Burke, J. (1986). *The day the universe changed* [television series]. London: BBC. Available from http://www.clearvue.com/productDetail.asp?objectID=11327.
CNN. (2004, January 30). *Carter slams Georgia's "evolution" proposal*. Available at http://www.cnn.com/2004/EDUCATION/01/30/georgia.evolution/
Cobern, W. W. (1994). Point: Belief, understanding, and the teaching of evolution. *Journal of Research in Science Teaching*, 31, 583–590.
Darwin, C. (1976). *The origin of species* (6th ed.). New York: Collier. (Original work published 1872)
Dawkins, R. (1998). *Unweaving the rainbow: Science, delusion, and the appetite for wonder*. New York: Houghton Mifflin.
Dawkins, R. (2003). You can't have it both ways: Irreconcilable differences? In P. Kurtz (Ed.), *Science and religion: Are they compatible?* (pp. 205–209). Amherst, NY: Prometheus Books.
Delpit, L. (1995). *Other people's children: Cultural conflict in the classroom*. New York: New Press.
Dennett, D. C. (1996). *Darwin's dangerous idea*. New York: Touchstone.
Dewey, J. (1916). *Democracy and education: An introduction to the philosophy of education*. New York: Macmillan.

Dewey, J. (1933). *How we think: A restatement of the relation of reflective thinking to the educative process.* New York: Heath.
Dewey, J. (1959). The child and the curriculum. In M. S. Dworkin (Ed.), *Dewey on education* (pp. 91–111). New York: Teachers College Press. (Original work published 1902)
Dott, R. H., Jr., & Batten, R. L. (1988). *Evolution of the Earth* (4th ed.). New York: McGraw-Hill.
Eldredge, N. (1982). *The monkey business: A scientist looks at creationism.* New York: Washington Square Press.
Eldredge, N. (2001). *The triumph of evolution and the failure of creationism.* New York: W. H. Freeman.
Good, R. (2005). *Scientific and religious habits of mind: Irreconcilable tensions in the curriculum.* New York: Peter Lang.
Gould, S. J. (1977). *Ever since Darwin.* New York: Norton.
Gould, S. J. (1983). Evolution as fact and theory. In *Hen's teeth and horse's toes* (pp. 253–262). New York: Norton.
Gould, S. J. (1999). *Rocks of ages: Science and religion in the fullness of life.* New York: Ballantine.
Haught, J. F. (2000). *God after Darwin: A theology of evolution.* Boulder, CO: Westview Press, Perseus Books.
Jackson, D. F., Doster, E. C., Meadows, L., & Wood, T. (1995). Hearts and minds in the science classroom: The education of a confirmed evolutionist. *Journal of Research in Science Teaching, 32,* 585–611.
Kohl, H. (1992, August). I won't learn from you! *Rethinking Schools, 7*(1), 1, 16-17, 19.
Lawson, A. E. (1995). *Science teaching and the development of thinking.* Belmont, CA: Wadsworth.
Matsumura, M. (2001). *Six significant court decisions regarding evolution/creation issues.* Berkeley, CA: National Center for Science Education. Available at http://www.ncseweb.org/resources/articles/620_seven_significant_court_decisi_12_7_2000.asp.
Menand, L. (2001). *The metaphysical club: A story of ideas in America.* New York: Farrar, Straus, and Giroux.
Miller, K. R. (1999). *Finding Darwin's God.* New York: Harper Collins.
Newport, F. (2004). *Third of Americans say evidence has supported Darwin's evolution theory; Almost half of Americans believe God created humans 10,000 years ago.* Princeton, NJ: Gallup News Service.
Nord, W. A. (1995). *Religion and American education: Rethinking a national dilemma.* Chapel Hill: University of North Carolina Press.
Palevitz, B. A. (2003). Science versus religion: A conversation with my students. In P. Kurtz (Ed.), *Science and religion: Are they compatible?* (pp. 171–179). Amherst, NY: Prometheus Books.
Polkinghorne, J. (1998). *Belief in God in an age of science.* New Haven: Yale University Press.
Pope John Paul II. (1996, October 22). *Message to the Pontifical Academy of Sciences.* Available at http://www.newadvent.org/library/docs_jp02tc.htm.
Prothero, D. R., & Dott, R. H., Jr. (2003). *Evolution of the Earth* (7th ed.). New York: McGraw-Hill.
Rosin, H. (2005, January/February). Beyond belief. *The Atlantic Monthly, 295*(1), 117–120.
Scott, E. C. (2000). *The creation/evolution continuum.* Berkeley, CA: National Center for Science Education.Available at: http://www.ncseweb.org/resources/articles/ 9213_the_ creationevolution_continu_12_7_2000.asp.
Skehan, J. W., & Nelson, C. E. (2000). *The creation controversy and the science classroom.* Arlington, VA: National Science Teachers Association Press.
Smith, M. U. (1994). Counterpoint: Belief, understanding, and the teaching of evolution. *Journal of Research in Science Teaching, 31,* 591–597.

Teaching for Understanding Rather than the Expectation of Belief

Leslie S. Jones

Recognizing the Resistance

During a graduate seminar in science education, my cohort of doctoral students was invited to discuss a situation that was perplexing the departmental faculty. Apparently, an undergraduate student majoring in science education had refused to give a series of high school biology lessons on evolution during his student teaching on the grounds that it was against his religion. His supervising teacher at the high school had written a negative evaluation and the department faculty was considering the possibility of withholding his licensure and certification. Most of us were older graduate students with diverse and considerable cumulative experience in education, and so the situation generated lively discussions and a wide range of opinions. I distinctly remember starting out on one side of this issue but changing my mind dramatically after talking with my colleagues and thinking about this situation.

Initially, I considered the refusal to address evolution an indefensible position for a biology teacher. If he did not believe in evolution, he could certainly have chosen another field for specialization. Since evolution is the backbone of biological science, it seemed incomprehensible to exclude it from the curriculum. However, as the situation festered over time and our faculty argued about what should be done, conversations with my advisor opened my eyes to other facets of the situation. I felt a growing empathy for the student as we watched how he was treated by his teachers on the faculty. I joined several of my contemporaries who were becoming increasingly critical of our program for letting this happen. Several serious questions were raised. How could he have gotten through all of that biology content coursework and feel that way about evolution? How did he go that far in his educational program without drawing attention to his opinion? What made him think that evolution was so incompatible with his religion? Most importantly, why was our department intent on treating him as a pariah rather than recognizing his religious rights and advising him about the situation?

I never met the young man or had the chance to get answers about his side of the story, but I realized that the institutional response was largely based on embarrassment, curricular rigidity, and a lack of concern for the person involved. The insensitive dismissal of the importance of religious values was disappointing, even though it was easy to understand why the department was concerned. The theory of evolution is the single most important concept developed in the human quest to understand the living world, and biology teachers should be expected to teach evolution. However, a critical aspect of science teacher education has to be preparation to teach the subject. My immediate suspicion was that the student teacher probably did not understand the theory of evolution. If nothing else, our science education program had certainly failed to address the subject at an appropriate stage in his education.

I silently cheered when I heard the faculty could not prohibit that young man's teaching certification. He was not to blame for the fact that the university faculty had failed to recognize the problem earlier. Several issues should have been resolved before he started his internship in the school system. Education professors should have known the students going through teacher certification well enough to help them resolve significant personal issues like this. In the end, we wondered what his future biology classes would be like without evolutionary content.

Prior to the debate over whether or not this person should be a certified biology teacher, I had never considered the possibility that people could understand biological science without recognizing that it is all based on an underlying evolutionary framework. I maintained my conviction that evolution should be taught as a central aspect of every life science curriculum, but this incident opened my eyes to the fact that some people seriously consider evolution to be incompatible with their religious convictions. At the time, few people in science education could have predicted how controversial this issue was about to become. I completed my doctoral program and started teaching college biology in 1997; just two years before the infamous decision of the Kansas School Board to remove evolution and related concepts from their state science standards brought this simmering issue to a rapid boil on a national scale.

As fate would have it, the first time I taught about evolution in a college biology class, I was reintroduced to the evolution/creationism controversy. As a science educator in the biology department of a Midwestern university, I taught large sections of required science courses for non-majors. At the mere mention of the word evolution on the first day of class, there was a palpable tension in the lecture hall. I had the distinct impression that some of these students were surprised to hear their biology professor even mention the subject. Since I knew that it would be unrealistic to even try to teach biology without depending on the explanatory power of evolutionary theory, I needed to find out what these students were thinking. At the end of that class session I asked each person to hand in a short, written statement answering the question: "How do *you* believe life began and humans came into existence, and why do *you* believe what *you* believe?"

The initial collection of responses to that question opened my eyes to the significance of what has become the most pervasive challenge in biological education. Students cited personal views that ranged from strict biblical creationism to complete acceptance of evolution. However, the distribution of opinions was heavily skewed toward creationist views. Even though some of the students indicated that their ideas were shaped by a combination of religious and scientific information, almost half of the class espoused strictly biblical perspectives on the topic of origins. Many of these creationist students made it clear that they had serious reservations about even being exposed to evolutionary science. They seemed convinced that evolution was incompatible with their religious convictions and would be presented as an alternative explanation. They justified their aversion to evolution using statements that exposed fundamental misunderstandings of the nature of science. Evolution was misconstrued in ways that showed they simply did not understand scientific explanations for the origin of the universe or the history of living organisms. Some of the most adamant creationist students used erroneous statements about evolution as the primary reasons they did not consider the scientific ideas to be valid or even worthy of consideration.

This was my first exposure to anything other than a public distillation of creationism, and I realized that teaching evolution to creationist students was going to be a serious challenge. I had always thought that the basic creationist objections to evolution were based on a perceived need to defend religious authority, but the words of my own students demonstrated that I really did not comprehend what most creationists were thinking. In general, the students' essays contained a mixture of defensive theology and anti-science rhetoric. I might have predicted the religious opinions, but I was not expecting their apparent distrust of scientific ideas. A substantial proportion of this student population was well-equipped with arguments that purposely attempted to discredit evolutionary theory. Since I felt that I could not afford to ignore the complexities of the attitudes these students brought to my classroom, I decided to address their reservations in the context of my presentation of evolutionary science. I sensed, after reading their personal opinions, that diffusing the controversy was a crucial step in preparing these students to learn about biological evolution.

Untangling the Complexity

Ever since students reacted to the word "evolution" in my first college biology course, teaching strategies that address creationist opposition have been an important part of my presentation of evolutionary theory. Even when I repeat a particular course year after year, I have students write these personal essays on their views of origins so that they take the time to consider and have a chance to express their opinions. Furthermore, I get a sense of the particular audience in every class. I continue to learn from student essays on "why they believe what they believe," allowing me to directly address more specific aspects of creationist reservations. Individuals articulate their ideas using different words, but there are fundamental variations on a limited number of emergent themes that are helpful in understanding how to defuse their apprehension about learning about this subject.

A recent move from my initial teaching position in Iowa to a university in rural South Georgia has extended my initial observations of creationism in the Midwest to a group of students in what is commonly known as the "Bible Belt Region." My current students have grown up in an extremely conservative populace where public rejection of evolution is widespread. I teach courses for a variety of students including: biology majors, non-majors in Liberal Arts classes, and education majors in required science content classes. It is interesting that students in this part of Georgia express views that cover the same spectrum of opinions as students in Iowa. There are stronger statements about their complete faith in God and Christian identities, but the nature of creationism seems very much the same. One interesting difference is that Georgia students admit that they really do not understand evolution because they have never been taught anything about it in school.

Teaching in Georgia has been a lesson in how both the religious background and the level of prior exposure to evolutionary science influences the formation of particular opinions. Evolution is so rarely mentioned in Georgia schools that these students actually tend to be somewhat curious to learn what it is. In general, they seem very confident of their religious convictions and not as concerned that science will ever change their minds about God. I suspect there is less antievolution rhetoric circulating because of the assumption that evolution would not be taught in the local public schools. (This is likely to change since the new state curriculum explicitly includes biological evolution despite an attempt by the state superintendent of education to have the word evolution removed from the Georgia Performance Standards.) Creationism is assumed, by default, to be the majority opinion, and some students are shocked that I would even ask how they "think life began and humans came into existence."

More than half of the students I surveyed in Iowa and close to a third of those I have taught in Georgia are not strict creationists. On the basis of well over 1,000 personal essays, it appears that some college students have no trouble accepting scientific explanations while maintaining loyalty to their religious beliefs. Some individuals combine religion and science by expressing a combined belief in a biblical account of creation as well as various aspects of evolutionary science. They seem to assimilate the two, accepting both religious and scientific sources of authority with emphasis on the importance of both types of information. Other religious students express this basic idea somewhat differently, as scientific theism. This position frames the acceptance of religion and science within the idea that God is ultimately responsible for what has happened and evolution is an explanation for how everything happened. This, more theologically liberal position comes from students who interpret scripture allegorically and feel no need to take the creation story literally. They accept aspects of evolution without giving up the idea that God played the most important role in the origin of the universe.

In both universities, only a very small number of students, usually less than five percent, think of origins exclusively in terms evolution. Of those who take exclusively scientific positions, either rationalism or an agnostic lack of inclination toward religion are given as their reasons. Rationalists talk about a need for evidence and how they are convinced that scientific data confirm evolutionary theory. People who are either agnostic or atheistic have no strong religious beliefs, no confidence or proof that God exists, nor a conviction that God does not exist. I suspect that these numbers might be particularly small because evolution has a negative connotation in both regions. In Iowa there is a perception of social pressure to be at least somewhat religious, and in the southern part of Georgia, people are expected to be very religious. In any case, teaching evolution to these students is, as they say, "preaching to the converted," and the challenge is in reaching students with an especially strong aversion to being told about the theory of evolution.

Among the students I have worked with in both Iowa and Georgia, religious creationism is expressed as a strict commitment to the idea that God created life in its present form as described in *Genesis*, the first chapter of Hebrew Scripture or the Old Testament of the Bible. This view is supported by three basic themes: belief in God, literal interpretation of the Bible, and loyalty to a particular religion. Avowed creationist students are adamant that God alone is responsible for the origin of the universe and everything happened precisely as described in the first chapter of Genesis. Their statements contain strong convictions that their faith in God is all of the proof they need. Many of them consider the Bible to be the inerrant word of God and therefore the most appropriate explanation of creation. They profess biblical literalism and commonly include scriptural quotations from both the New and Old Testament to support their reasoning. Some creationists base their stance on loyalty to a particular religious denomination. These religious students discuss a belief in creationism to be part of the what they perceive as the expectations of their church since the idea was a dominant aspect of what they were taught as part of their religious education. Statements of religious/denominational loyalty come from members of a variety of churches and are clearly not confined to members of conservative Protestant congregations.

One misconception educators hold about creationism is the idea that resistance to evolution is confined to fundamentalist and evangelical Christians. Not only can Methodists, Lutherans, Presbyterians, and even Episcopalian Protestants be strict creationists, but Roman Catholic students articulate some of the strongest views on this subject. Roman Catholic students, especially in Iowa, talk about how they have learned that evolution goes against the Bible from priests, in parochial schools, and in catechism classes. This prevalence of fervent creationist beliefs among Roman Catholics and their tendency to view evolution as being incompatible with their religion reveals that members of this mainstream religion see a basic conflict between evolution and their church, despite the evidence that such a theological schism does not officially exist.

For the last fifty years, the leaders of the Roman Catholic Church have clearly articulated the idea that evolution is a scientific theory that is compatible with the basic tenets of their faith. Pope Pius XII stated in 1950, as part of his encyclical *Humani Generis*, that the church had no opposition to evolution or the idea that the human body originated from preexistent living matter, as long as God is recognized for creating the spiritual soul. Pope John Paul II further elaborated the Church's position when he stated that "New knowledge has led to the recognition of the theory of evolution as something more than just a hypothesis" and that discoveries had provided "significant argument in favor of this theory" (Christus Rex Information Service, 1996, p.2).

Many Roman Catholics are apparently unaware that the *Catechism of the Catholic Church* (United States Catholic Conferences, 1994) teaches as part of the official Catholic doctrine that God created life, but not specifically how it was done. In chapter one, "The Profession of Faith," there are detailed descriptions of creation, including passage 341 on "the beauty of the universe," which states that "The order and harmony of the created world results from the diversity of beings and from the relationships that exist among them. Man discovers them progressively as the laws of nature. They call forth the admiration of scholars. The beauty of creation reflects the infinite beauty of the Creator and ought to inspire the respect and submission of man's will" (p. 88). The Vatican has publicly supported the teaching of evolution by stating that "intelligent design" is not science and only creates confusion when it is presented as such.

The intensity of the evolution/creationism controversy is, however, not rooted strictly in the defense of religious ideas. I realized this when it became obvious that creationist students were talking about more than religion in their essays. Even though I ask "what you believe and why you believe it," creationist students often take the initiative to go beyond their beliefs to make it clear that they are convinced that evolution is incompatible with their faith in God. The tone of these comments indicates the perceived need to defend their faith and the presumption that religion is going to be belittled or challenged by scientific ideas. In what is best described as skepticism about science, a variety of comments openly display a lack of confidence in the evidence used to support the theory. Furthermore, some students apparently doubt the validity of the empirical verification and question the intentions and honesty of the scientific community in ways that show obvious cynicism about science. This mistrust of science has been the most significant revelation in all of the students' testimony. Striking similarity between their statements and creationist rhetoric in the public arena demonstrates that these students have learned and internalized this anti-science dogma from political groups that challenge the teaching of evolution in public schools.

Addressing the Controversy

In the context of the current controversy, teaching evolution has to be much more than merely presenting an accurate portrayal of the biological basis of the theory. Even though the evolution/creationism controversy is phrased as a defense of religion, political agendas are surreptitiously hidden behind the idea of defending faith. The comments of students in my courses make it clear how the current iteration of creationism has become an open challenge to science and much more than just a religious issue. Therefore, effective teaching of evolution depends on initially defusing the problem with a strategy that addresses both theological and antiscientific dimensions of the issue. Without addressing the prevailing misinformation that apparently receives greater distribution than the actual scientific ideas, we have little chance of teaching sound science to creationist students.

Many students are not ready or willing to learn about evolution when they enter our classrooms. In fact, some will be honest about the fact that they intend to go through the motions of doing whatever it takes to earn passing grades while working hard to resist the supposedly immoral influences of science. Because of their reticence, my actions in the classroom have been shaped by a need to synthesize a methodology for teaching evolution that addresses many contentious viewpoints. Before I attempt to introduce the subject, I try to disarm the current controversy and diminish overt resistance to the theory. Given the current distorted portrayal of evolution in the mass media, this means that in general science courses I often spend as much time addressing the ongoing social controversy as I actually spend teaching about the scientific theory of evolution. Fortunately in courses that focus on evolution I can spend proportionally more time clarifying the scientific ideas.

In every course, I make it clear that the theory of evolution is going to be an important instructional theme. A basic appreciation of this topic is one of the most critical elements of scientific literacy and since it is being neglected elsewhere, there is good reason for me to address this as a central topic. I want students to understand how evolution serves Biology as the only reasonable analytic framework used to interpret the history of life. I emphasize that it would be impossible to describe scientific interpretations of natural phenomena without using evolutionary concepts. I explain why it is impossible to teach biological science without employing the single, underlying, explanatory framework of the discipline.

My first instructional activity is to assure religious students that their personal belief systems are safe in my classroom. This is not simply a matter of making a blanket statement; it is a matter of developing trust and continually reinforcing that promise. As much as I would like to start courses by setting down evolution as the conceptual framework for biological science, I usually give students time to form an impression of me as a teacher. The length of time varies from two days in short summer courses to three weeks in full semester courses. I acknowledge that I realize many students are uncomfortable learning about evolution because they have been led to believe that this scientific theory is antireligious and that the people who teach evolution are trying to lead students away from the belief in God. I prompt students to consider the nature of different types of information and do not debate the relative merits of science or religion as valid sources of knowledge. Students are told that I expect them to learn about and understand biological evolution, but they are not obligated to "believe in" evolution or give up any of their religious convictions.

I encourage the students in my courses to realize that, as educated people, any judgment they make about biological evolution should be based on understanding of what the theory actually is rather than on fallacies that are often circulated about it. I point out that it is curious that the large body of mythology related to evolution apparently gets wider recognition in the popular press and within the general public than accurate explanations of this important scientific theory. By definition, creation myths are stories concerned with the formation of the world and its inhabitants, so perhaps it should be no surprise that the theory of evolution has been surrounded by so many "beliefs whose truthfulness is accepted uncritically." People often believe many of the myths associated with evolution because they have never heard an adequate presentation of evolution. If evolution is not explained in science classes, it is hardly surprising that citizens are not able to distinguish the truth from the fiction that circulates about evolutionary theory.

Many creationist students have been led to believe that evolution contradicts the Bible and discredits scriptural accounts of creation. Since there is nothing in evolutionary theory that directly refutes the accounts of human origins in the Bible, I invite them to satisfy themselves by looking critically for this in my explanations and the chapters in their textbooks. I realize that for those who believe that the Bible is the inerrant word of God, the fact that evolution is not described in the Bible can be problematic. I suggest that it is important to consider the fact that in the period of time that the Bible was written, people would not have had the necessary language to explain or even discuss evolution. Since Biblical and scientific explanations of human origins have come about in very different periods of history, they are based on very different types of knowledge.

One critical idea that must be deconstructed is the notion that all scientists are atheists. In Iowa, I was able to adopt what I considered to be an appropriately neutral stance and keep my personal beliefs out of class discussions. In Georgia, it became obvious that maintaining that level of privacy leads some students to automatically assume that I am an atheist. After being pointedly asked if I am an atheist, I realized that some students wondered why else would I ask them to explain why they believe in the Bible. Because it matters to my students, I have learned to be more comfortable sharing the fact that science has never seemed incompatible with my own belief in God. This demonstrates how important it is to recognize and make concessions for regional differences in expected behavior.

Atheism actually presents a very interesting semantic issue. From the standpoint that evolution is not theistic or based on a belief in God it is, in one sense of the word, atheistic. On the other hand, evolution does not deny the existence of God, and so it is not atheistic in the way most people think. Evolution is, in fact, absolutely neutral on the issue. Even those who are not strict creationists have commonly made the default assumption that the scientific explanation automatically contradicts any theistic premise. I make it clear that scientists have no atheistic agenda or intention of proving that God does or does not exist. Questions regarding the existence of God are not part of the material dimensions of the natural world. Proving or disproving issues based in faith are completely outside of accepted methods and processes that are the practice of science. Science is a way to assemble naturalistic explanations based on empirical evidence. The fact that scientists do not make reference to God when explaining the history of life is analogous to the way that God would not be credited or blamed for something like the proliferation tabloids and other aspects of the mass media.

As the controversy over evolution increases, civil conversations about religion and science seem to be conspicuously rare. My own understanding was critically influenced by continuous conversations with two close friends. One is a deeply religious Protestant biology teacher who was initially afraid that teaching evolution would take her students away from God. She was able to clarify for me why students feared science and what aspects were most commonly misunderstood. In return, I was able to elucidate evolutionary theory in a way that demonstrated the fallacies of some of the things she had been told. The other friend is a Roman Catholic who has just finished a Masters of Divinity. While she was in graduate school, she shared valuable references and explained how distortions of theology were exacerbating the controversy. We exchanged ideas in many conversations as we both wrote papers addressing this issue in our respective academic disciplines.

Realizing that discussions with these and other religious people have helped me understand the relationship between religious and scientific thinking, I decided that it might help students to hear a scientist hold a conversation with a member of the clergy. Two years ago, a friend and I videotaped a conversation called, "The Preacher and the Teacher." This is part of the instructional CD in the lab manual for a large introductory non-majors course called "The Evolution and Diversity of Life" at Valdosta State University. The pastor, Dr. Milton West, is the senior minister at the First Christian Church across the street from the university and literally lives down the street in my neighborhood. On the CD, we ask each other personal questions that demonstrate the ways we make sense of the existing furor over the teaching of evolution in the context of our religious backgrounds and beliefs.

Another way we have addressed the creationist reticence among our students is by printing an essay called "Myths and Truths about the Theory of Evolution" as an appendix in the same lab manual. I wrote this piece to assure students in my classes that I had no intention of challenging their religious beliefs as part of our lessons on evolution. In the paper I acknowledge that I know students can be uncomfortable learning about evolution if they have been led to believe that this scientific idea is antireligious and the people who teach evolution are trying to lead students away from the belief in God. I urge them to be certain that any judgment they make about biological evolution is based on an understanding of what the theory actually is, rather than the fallacies that are often circulated about it. I clarify the seven myths and the three truths listed below:

Myths
- Evolution Contradicts the Bible, Discrediting Genesis
- Evolution Is a Deliberate Test of Faith in God
- Evolution Explains the Origin of Life
- Evolution Is Part of Modern Social Decadence
- All Scientists Are Atheists & Try to Prove that God Is Not Responsible
- Evolution Is Incompatible with & Attempts to Replace Religion
- Accepting Evolutionary Explanations is Sacrilegious

Truths
- Evolution is a Theory
- Evolution Explains the History of Living Organisms
- All of Biological Science Is Based on the Theory of Evolution

Interestingly, but hardly surprising, these students (who are taking the course to satisfy a science requirement) do not take the time to look at either the video or the appendix unless they are prompted by an assignment or the lure of extra-credit for submitting an anonymous reaction. At the end of one semester, I asked the 100+ students in my lecture section how many had noticed either the video or the appendix and only two or three students raised their hands. During the next semester when I was not teaching either a lecture or lab section of the course, each graduate teaching assistant offered one point of extra credit for students who were willing to evaluate the two course supplements. Their comments ranged from individuals who openly wondered why the issue received so much attention ("It is no big deal to me") to those who felt defensive ("This does not change my belief in God"). The vast majority of the students stated that they learned things that they had not realized about evolution. So, instructors continue assign the CD video and written appendix as part of the course content, but unless there is accountability via a quiz or written assignment, we all know that students are unlikely to actually take the time to do this.

One of the other efforts I make to address the evolution/creationism controversy takes place literally outside of my own classes. Our new science building has a number of impressive glass-enclosed cases in an atrium that has a very high volume of student traffic. I reorganized most of these displays as self-standing visual lessons on evolutionary biology. Some focus on the diversity within a group of organisms such as sponges, corals, and sea turtles thanks to an extensive collection of marine specimens left by a retired professor. Other windows explain evolutionary themes with printed backdrops clarifying scientific concepts such as interspecific/intraspecific variation or the substrate-dependent adaptations of some marine organisms.

The most explicit evolution lessons in these display cases use our departmental collection of mammalian skulls. One window contains twenty-six different specimens and a challenge to guess the animal and what it eats. The key with the answers is located out of view in the next display. We remove the key and use this as a lab activity in the non-majors biodiversity lab. The other and most specific depiction of evolution includes plastic casts of various ancestral hominids that were originally just lined up with tags containing their scientific names. After securing permission to take the few primate skulls out of the mammology storage closet and purchase an expensive assortment of other primates, including all of the great apes, I made a backdrop with colorful placards clarifying some of the "Myths and Truths" about human evolution. I honestly wondered about the possibility of vandalism since this area is considered to be such a bastion of creationism, but nothing has happened in the three years since the exhibit was constructed. I can watch people stare at the display from inside my classroom and I see parents with children and school teachers with their classes study the explanations.

Telling the Story

In spite of the fact that evolution is such an important theme in biological science, it can be one of the more difficult concepts to understand. The strength and the brilliance of the theory is that it is holistic and presents a unified explanation of the living world that amounts to a reality that is far more than just the sum of all of its parts. Evolutionary processes are the basis of scientific explanations of the history of life on earth. The theory is based on a wide variety of evidence, and the nature of some of that evidence is what makes evolution harder for nonscientists to recognize. Most of biology is relatively concrete because the studies examine existing physical entities such as the cell or the immediate outcome of a particular phenomenon such as the molecular basis of heredity. Historical patterns of evolution, on the other hand, are basically inferential and abstract. For the most part, it is the consequences of evolution that are observed when describing life's history since the timing of most processes preclude direct observation. The presentation of a well-constructed explanation of evolution is crucial and misunderstandings are propagated when instruction is not effective.

One of the greatest mistakes in teaching evolution is the choice of the voyage of the Beagle as the opening for a unit of study. In spite of the fact that many textbooks highlight Darwin's journey around the world, this topic does not do justice to the history of the theory. It is also not particularly effective with creationists for two reasons. First, Charles Darwin has been demonized by the mistaken assumption that he caused the whole problem by "inventing" evolution. The story of his trip reinforces the assumption that he is to blame and emphasizing his journey also completely detracts from the accurate history. Evolution was recognized long before Darwin and it would be much more accurate to disseminate the fact that evolution is described in the writings of the ancient Greek philosophers such as Anaximander around 600 BCE. Second, for the novice science student, the Galapagos finches all look basically alike and do little to make a convincing case for selective adaptation. At a later stage in biological education, this can be a wonderful example, but there are much better ways to build a convincing presentation of evolution for an audience of skeptics. A little known aspect of history is the fact that these birds made little impression on Darwin until ornithologist John Gould pointed out that they were all finches and showed evidence of distinct adaptations to different feeding strategies.

The other typical centerpiece of short units on evolution, natural selection, is also a questionable choice for immediate study. Even though natural selection can serve as an explanation for genetic change over time, it is not convincing when isolated from the ecological circumstances that cause it to take place. Natural selection is frequently oversimplified and misconstrued as a dramatic process of speciation due to "survival of the fittest" organisms. Natural selection more accurately refers to the historical pattern of gradual proliferation within a population of some variants of a particular species due to inherited characteristics that lead to higher levels of reproductive success. Emphasis on mere survival is incorrect because it is the genes coding for favorable heritable characteristics that are actually proliferating in the process, and life span is irrelevant.

There is little question that the vast array of evolutionary information must be reduced for presentation, but there are much better choices than Darwin's voyage and natural selection. At the most basic level, biological evolution is the sum of the changes in populations of living organisms. Evolution accounts for the similarities as well as the differences among everything from bacteria to humans. While anyone can see the differences between simple microbes and such a complex animal species, it is evolution that accounts for how the same DNA molecules code for heredity in every living thing. Studies of the characteristics of the vast array of living species provide strong visual evidence of the similarities that are due to the historic relationships between species. Most recently molecular evidence gained from DNA studies has refined our understanding of the genealogical relationships among organisms and provided impressive corroboration for the relationships inferred from patterns of morphological similarities and differences.

Biological diversity is a preferable choice for demonstrating the strength of evolution in the contentious atmosphere created by the evolution/creationism controversy. The description of the historical changes and increasing complexity of living organisms is a concrete phenomenon that is accessible to students without scientific backgrounds. Understanding biodiversity begins with the process of grouping organisms on the basis of obvious physical similarities. Even though Linnaeus, the father of taxonomy, did not believe in evolution, the system he started is now centered on categories based on the presumption of common ancestry. Since the physical resemblance among grouped organisms is evident and the obvious differences correspond with categorical divisions, these classification schemes generally make sense to novice observers. Biodiversity models evolutionary theory because explanations for homologous morphological similarity are based on common origins. Even for students without sophisticated understanding of genetics, their intuitive understanding of heredity can be used to make a case for scientific assumptions that similar characteristics such as DNA, cells, and metabolic processes are indications of common descent. Starting with information about the evidence for common ancestry makes it much easier to subsequently introduce mechanisms of evolutionary divergence.

For creationist students, one of the least provocative examples of evolution is an explanation of the ancestral relationships among photosynthetic organisms. The question of human origins seems to be the most controversial aspect of biological evolution, and since plants have little recent connection to human ancestors, the subject is met with less resistance. Direct observation with and without microscopes allows students to see that that organisms from cyanobacteria to flowering angiosperms are all green in color. Other uncomplicated classroom lab activities, such as the use of paper chromatography, demonstrate that similar pigments are present in photosynthetic species. Lecture explanations help students build an understanding of the evidence that leads scientists to assume that common ancestry is the reason that chlorophyll is involved in the conversion of radiant to chemical energy in so many different organisms. The fossil record of this group is also particularly convincing since the living stromatolites in Western Australia document how cyanobacteria today can be compared to comparable fossil structures to gain important clues about the earliest organisms of this type. After building a case for the unthreatening subject of plant evolution, the parallel presentation of animal evolution seems more credible and less provocative.

A very common misrepresentation is the idea that evolution explains the origin of life. The popular press commonly misrepresents evolution as an explanation for the origin of life, even though scientists do not make that claim. In my classes, I teach that scientists would love to know how life actually began, but I confess that the closest we have come is to speculate on how life *might* have begun. One of the greatest quests of science has been its attempts to unravel the mysterious transition from inert chemicals on early earth to the original living organisms. I always screen this explanation in any textbook I consider and find that reputable college texts accurately represent the fact that we have never been able to explain the critical moment when life began. Attempts to denigrate evolution misrepresent the famous Miller–Urey experiment (depicted by diagrams of glass chambers containing chemicals being subjected to electrical energy) as a scientific claim to generate life. Scientists know that this study only demonstrated the abiotic synthesis of organic compounds, not the formation of living organisms.

It gives me great pleasure to agree with the creationist declaration that "evolution is just a theory." In that a theory is, by definition, a rational group of tested general principles of explanation for a class of phenomena, the theory of evolution is certainly a theory. The abundance of supporting evidence and the absence of major contradictions to evolution are the reason that it is considered to be such a well-established proposition. I clarify the specific meaning of terminology related to the nature of scientific processes so that it is clear that evolution is not merely a hypothesis or provisional conjecture used to guide investigation. I emphasize that the theory is not speculative because it has been repeatedly verified by evidence collected by the scientific community. Scientists accept the explanatory value of evolution as fact and most of us do not consider our confidence in the theory to be a "belief" because that term connotes confidence in the truth of something that cannot be subjected to rigorous testing or proof. Evolution does consider common ancestry the reason that humans are so similar to other living primates, but evolution does not propose that we descended from modern apes. The fact that monkeys in the zoo do not morph or change into humans does not prove evolution is not true, as many students claim, because they are not our direct ancestors.

Responding without Debate

Whether or not there has ever been any local challenge to the teaching of biological evolution, personal views are shaped by experiences prior to the age at which science classes begin to formalize the presentation of natural history. Students in any biology class may or may not even be conscious of their own feelings about the origin of life and/or the way humans came into existence. Unless they have a strong inclination to believe either scientific explanations or scriptural accounts, college students may never have actually considered their own personal views. It is often challenging for them to begin to articulate their views, but they appreciate having the chance to do so. From an instructional standpoint, students who either consider evolution to be the preferable explanation for human origins or have no problem reconciling biblical and scientific ideas benefit from the presentation of biological evolution because even they often hold misconceptions as to what the theory actually means. It is, however, the students with strong creationist beliefs, especially those with preexisting aversions to evolutionary theory, who provide the greatest challenge and are actually the focus of many of the strategies I employ in my classes.

Given the importance of addressing the fears of creationist students, it has become crucial to discuss the relationships between religion and science. There is an unfortunate tendency for creationists to assume that evolution is incompatible with religion. This fallacy demonstrates their unwillingness or inability to recognize the fundamental distinctions between religion and science. I have developed a variety of inquiry-oriented lessons that center on the nature of science and emphasize the fact that science is the systematic study of the natural world. Students are encouraged to observe the consequences of natural selection in a lab that uses live minnows to show both adaptive behavioral responses and cryptic coloration that impact survival and chances of reproductive success. I emphasize that scientific knowledge is based on materialistic explanations which are part of a view that matter and its motions constitute the physical universe. I want students to recognize that religion is a completely different worldview because religious explanations are necessarily immaterial or incorporeal which means they are spiritual and not based on physical worldly matters. On the basis of the reflections in student portfolios and conversations inside and outside of the classrooms, students recognize that I am not trying to contravene their religious views.

I talk about religion from a philosophical perspective without teaching or advocating any particular tradition. It is amusing to see the quizzical looks on some faces when I discuss Christianity as if it is not normative. Those students apparently think that most people in the world are Christians. I explain why it is important to remember the importance of religious pluralism and respect everyone's right to make choices about religious beliefs. Somehow the idea of individual rights has been lost in both the creationist movement that opposes the teaching of evolution and a conservative political climate that seems to have lost sight of the reasons for the separation of church and state.

Several of my courses serve education majors who plan to enter the teaching profession, so there is good reason to help them recognize the sociopolitical dimensions of this issue. There are usually several students that make it clear that they think "both sides of the issue" should be presented, so the way I diminish this element of creationist resistance to evolution is by making it clear that while biological evolution is not the only explanation of origins, it is the only scientific explanation of human origins. Political challenges, especially from the recent intelligent design movement, create the false impression that scientists do not agree about the broad outlines of the history of living organisms. Students are invited to explore the intersections of science and spirituality in their own lives, but I make it very clear that I will not give equal time to the presentation of creation science or intelligent design because these are religious ideas and spiritual interpretations of scientific topics.

Given that I know many creationists assume scientists never think about religion, I share the following quotations attributed to two well-known scientists. Galileo is associated with the statement, "The Bible tells us how to go to heaven, not how the heavens go," even though he was apparently attributing the statement to Cardinal Cesare Baronius (1538-1607). Albert Einstein once stated that "Science without religion is lame, religion without science is blind." Both of these comments make powerful talking points when I speak to groups outside of my own classes.

I have come to the conclusion that personal belief systems are constructed under a variety of influences coming from both the public and private aspects of peoples' lives. Each person develops an individualized perspective that is the product of the intersection of intrinsic personality characteristics that are seriously modified by extrinsic social influences. The outside factors differ dramatically depending on the regional/cultural setting and the combination of Judeo-Christian creationism and exposure to scientific evolutionary theory. I developed the following schematic diagram (Fig. 12.1) to help students recognize the parallelism between scientific and religious influences on our views of human origins and to demonstrate that there is actually no monolithic concept or "epistemology of origins."

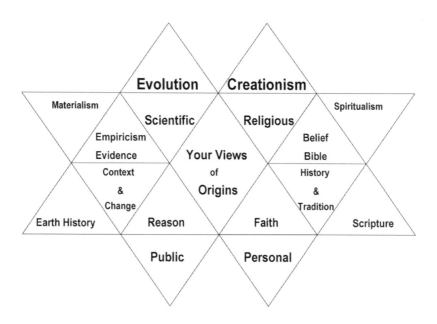

Figure 12.1 Parallel Influences on Personal Epistemologies

Many people are able to reconcile science and religion by assimilating the different ideas into a personal viewpoint that is informed by both philosophical doctrines. Other people accommodate religion and science as different realms of knowledge and take important ideas from each one. Unfortunately, some people fear that it is sacrilegious to accept evolutionary explanations. Certainly religious opinions vary and heresy depends on the degree of personal orthodoxy and particular doctrines of the religion someone practices. I urge students to be sure they understand what evolution really is, before they assume that evolution is strongly at variance with the important tenets of their faith. I just hope that most students leave my classroom with a reasonable level of evolutionary literacy.

If there is a lesson we can learn from the stubborn persistence of the evolution–creationism controversy, it is that teaching evolution involves far more than presenting biological explanations for the history of living organisms. Regardless of who is responsible for the controversy, and I do think that scientific arrogance is partially to blame; the scientific community can no longer afford to ignore the fact that biological education is being crippled by religious aversion to evolution. Important issues that impact receptivity to classroom presentations are often overlooked out of the failure to comprehend just how fundamentally incongruent this scientific idea can seem with the religious views of some people. The tendency to ignore religious objections on the grounds that they have no place in the public realm of science education has only strengthened resistance to this important biological premise.

The scientific community in the United States is only beginning to realize how serious this problem has become. Most college or university faculty members are not likely to feel directly threatened or worry that evolution will ever be prohibited in their classrooms. Students rarely challenge college faculty members directly, and their parents are not going to menace our administrations with threats of political consequences. While it is not a good excuse, it is also important to recognize that part of the enculturation we experience entering the profession stresses that scientific methods depend on objectivity and the products of science are supposed to be value-free knowledge. Even those scientists who recognize the social impacts of science are unlikely to be comfortable stepping out of the disciplinary frameworks even to defend evolution. There are a few activist scientists who are bold enough to "go public" and defend evolution in the media. It is usually the same dedicated individuals who are on the frontlines, taking the time to be involved in legislative battles and legal challenges. The problem is that the quality of biology instruction at precollegiate levels has already been impacted by the erratic and often minimal attention evolution gets. Therefore, these terminal science courses students take at our colleges and universities may be a chance to support the teaching of biological evolution by enhancing the scientific literacy of the general public.

References

Christus Rex Information Service (1996). News from the Holy See. Available: http://www.christusrex.org/www1/pope/vise10-23-96.html. Retrieved October 2001.

United States Catholic Conferences, Inc. (1994). *Catechism of the Catholic Church as translated from the Latin text*. Rome: Libreria Edtrice Vaticana.

13

Teaching about Origins in Science: Where Now?

Michael J. Reiss

As the earlier chapters in this book have made clear, teaching origins in science is an issue about which many people have strong views, and this is hardly surprising. There may not be a necessary, unidirectional, logical connection between where we came from, who we are, and what we should become, but the three questions relate to one another in most people's minds. The aim of this chapter is to draw together the threads of the book, clarify areas of agreement and areas of disagreement and suggest possible ways forward. I organize the chapter around three main themes: teaching about the nature of knowledge, controversial issues, and matters of personal significance.

Teaching about the Nature of Knowledge

In many countries, school science curricula have come in recent years to include more about the nature of science. Understanding the nature of science is a subset of the understanding of the nature of knowledge. Understanding the nature of knowledge lies at the heart of education, since, whatever the overall aims of education—whether the promotion of autonomy, well-being, justice, or otherwise (Marples, 1999; Reiss, 2007)—students need to know about the nature of knowledge. If they don't, how can they evaluate truth or other claims?

I would argue that helping students at any stage of education to understand the nature and scope of science (understood as the natural sciences) can help in the teaching of evolution to students who are predisposed not to accept the scientific understanding of it. The phrase "the nature of science" is used as shorthand for something like "how science is done and what sorts of things scientists work on." It therefore contains two elements: the practice of doing science and the knowledge that results from the process.

It is difficult to come up with a definitive answer to the question "What do scientists study?" Certain things clearly fall under the domain of science—the nature of electricity, the arrangement of atoms into molecules, and human physiology, to give three examples. However, what about the origin of the universe, the behavior of people in society, decisions about whether we should build nuclear power plants or go for wind power, the appreciation of music and the nature of love, for example? Do these fall under the domain of science? A small proportion of people, including a few prominent scientists, would not only argue "yes" but maintain that all meaningful questions fall within the domain of science; hence, use of the term "scientism" to refer, pejoratively, to the view that science can provide sufficient explanations for everything.

However, it can be argued hold that science is but one form of knowledge and that other forms of knowledge complement science. This way of thinking means that the origin of the universe is also a philosophical or religious question—or simply unknowable; the behavior of people in society requires knowledge of the social sciences (including psychology and sociology) rather than only of the natural sciences; whether we should go for nuclear or wind power is partly a scientific issue but also requires an understanding of economics, risk, and politics; the appreciation of music and the nature of love, while clearly having something to do with our perceptual apparatuses and our evolutionary history, cannot be reduced to science (Reiss, 2005).

While historians tell us that what scientists study changes over time, there are reasonable consistencies. First of all, science is concerned with the natural world and with certain elements of the manufactured world so that, for example, the laws of gravity apply as much to artificial satellites as they do to planets and stars. Secondly, science is concerned with how things are rather than with how they should be. So there is a science of gunpowder and *in vitro* fertilization without science telling us whether warfare and test-tube births are good or bad.

How is Science Done?

If it is difficult to come up with a definitive answer to the question "What do scientists study?" it is even more difficult to come up with a clear-cut answer to the question, "How is science done?" Robert Merton characterized science as open-minded, universalist, disinterested, and communal (Merton, 1973). For Merton, science is a group activity: even though certain scientists work on their own, all scientists contribute to a single body of knowledge accepted by the community of scientists. There are certain parallels here with art, literature, and music. After all, Leonardo da Vinci, Michelangelo, and Raphael all contributed to Renaissance art. But while it makes no sense to try to combine their paintings, science is largely about combining the contributions of many different scientists to produce an overall coherent model of one aspect of reality. In this sense, science is disinterested; in *this* sense it is (or should be) impersonal.

Of course, individual scientists are passionate about their work and are often slow to accept that their cherished ideas are wrong. However, science itself is not persuaded by such partiality. While there may be controversy about whether the poems of Ezra Pound or T. S. Eliot are the greater (and the question is pretty meaningless anyway), time invariably shows which of two alternative scientific theories is nearer the truth. For this reason, scientists are well advised to retain "open mindedness," always being prepared to change their views in the light of new evidence or better explanatory theories, and science itself advances over time. As a result, while some scientific knowledge ("frontier science") is contentious, much scientific knowledge can confidently be relied on: it is relatively certain.

Karl Popper emphasized the falsifiability of scientific theories (Popper, 1934/1972): unless you can imagine collecting data that would allow you to refute a theory, the theory isn't scientific. The same applies to scientific hypotheses. So, iconically, the hypothesis "All swans are white" is scientific because we can imagine finding a bird that is manifestly a swan (in terms of its appearance and behavior) but is not white. Indeed, this is precisely what happened when early white explorers returned from Australia with tales of black swans.

Popper's ideas easily give rise to a view of science in which knowledge accumulates over time as new theories are proposed and new data is collected to discriminate between conflicting theories. Much school experimentation in science is Popperian in essence: we see a rainbow and hypothesize that white light is split up into light of different colors as it is refracted through a transparent medium (water droplets); we test this by attempting to refract white light through a glass prism; we find the same colors of the rainbow are produced and our hypothesis is confirmed. Until some new evidence falsifies it, we accept it.

There is much of value in the work of Thomas Merton and Karl Popper, but most academics in the field would argue that there is more to the nature of science. Thomas Kuhn made a number of seminal contributions but he is most remembered nowadays by his argument that while the Popperian account of science holds well during periods of *normal science* when a single paradigm holds sway, such as the Ptolemaic model of the structure of the solar system (in which the Earth is at the centre) or the Newtonian understanding of motion and gravity, but it breaks down when a scientific *crisis* occurs (Kuhn, 1970). At the time of such a crisis, a scientific revolution happens, during which a new paradigm, such as the Copernican model of the structure of the solar system or Einstein's theory of relativity, begins to replace the previously accepted paradigm. The central point is that the change of allegiance from scientists believing in one paradigm to their believing in another cannot, Kuhn argues, be fully explained by the Popperian account of falsifiability.

Kuhn likens the switch from one paradigm to another to a gestalt switch (when we suddenly see something in a new way) or even a religious conversion. As Alan Chalmers puts it:

> There will be no purely logical argument that demonstrates the superiority of one paradigm over another and that thereby compels a rational scientist to make the change. One reason why no such demonstration is possible is the fact that a variety of factors are involved in a scientist's judgment of the merits of a scientific theory. An individual scientist's decision will depend on the priority he or she gives to the various factors. The factors will include such things as simplicity, the connection with some pressing social need, the ability to solve some specified kind of problem, and so on. Thus one scientist might be attracted to the Copernican theory because of the simplicity of certain mathematical features of it. Another might be attracted to it because in it there is the possibility of calendar reform. A third might have been deterred from adopting the

Copernican theory because of an involvement with terrestrial mechanics and an awareness of the problems that the Copernican theory posed for it.

(Chalmers, 1999: 115–16)

Kuhn also argued that scientific knowledge is validated by its acceptance in a community of scientists. Often scientists change their views as new evidence persuades them that a previously held theory is wrong. But sometimes they cling obstinately to cherished theories. In such cases, if these scientists are powerful (e.g. by controlling what papers get published in the most prestigious journals), scientific progress may be impeded—until the scientists in question retire or die.

Teaching about Religious Knowledge in Science Classes

Whether or not it is appropriate to teach students in science classes about the nature of religious knowledge, as well as the nature of scientific knowledge, will depend on the country, the year, the school and even the teacher. The strongest argument, in my view, for teaching anything about religion in a science class, whether at school, college or university, is if it helps students better to understand science. I am not here, therefore, addressing the more general question as to whether students should be taught in schools or college about religion for other reasons (e.g. that students should be introduced to the various domains of knowledge, one of which is religious knowledge with its distinctive claims, or because teaching about religion helps one better understand history or morality).

Teaching about aspects of religion in science classes could potentially help students better understand the strengths and limitations of the ways in which science is undertaken, the nature of truth claims in science, and the importance of social contexts for science. However, there are also reasons to be cautious before teaching about aspects of religion in science classes (Reiss, 1992). For example, science teachers might feel that they simply don't have the expertise to teach effectively about such matters, that these matters are better dealt with elsewhere in the curriculum, or that it is impossible to teach objectively about such matters so that one risks indoctrinating one's students either into or away from a religious faith.

However, what I have found to be of worth in science classes with undergraduates training to be science teachers is, when teaching about the nature of science, to get them to think about the *relationship* between scientific knowledge and religious knowledge. What seems to work well is to ask students, either on their own or in pairs, to illustrate this by means of a drawing, and then for all of us in the class to discuss the various drawings that result. See, for example, the hypothetical representation in Figure 13.1. A person producing the representation in Figure 13.1 sees both religious and scientific knowledge as existing but envisages the scope of religious knowledge as being smaller than that of scientific knowledge and of there being no overlap between the two.

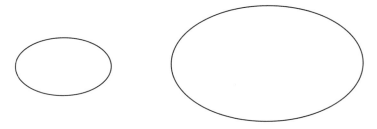

Figure 13.1 Hypothetical representation of how someone might see the relationship between religious knowledge (left) and scientific knowledge (right).

A number of authors in this book favor a clear-cut distinction between religious and scientific knowledge, along the lines of that defended by Gould (1999). There are a number of advantages to such a position. For example, it allows a person with a strong religious belief who might otherwise be troubled by certain aspects of science to avoid possible conflict (and vice versa) and it provides an epistemological justification for why religious matters should not be examined in science classes, which is useful in a country such as the US that prohibits the teaching of religion in schools.

However, there are many for whom scientific knowledge and religious knowledge are not distinct. At one end are those who draw religious knowledge as being much smaller than scientific knowledge and wholly or partly contained within it; at the other are those whose worldview is predominantly religious. In between are a whole range of portrayals of the ways in which science and religion can relate (c.f. Barbour, 1990). Understandings of the relationships between science and religion vary greatly, at least in part because of considerable variation in how people conceptualize both science and religion (Brooke, 1991). The visual metaphor in Figure 13.1 can be taken too far, but it can serve as a useful heuristic device.

Teaching about Controversial Issues

A widely accepted definition of a controversial issue is that "A matter is controversial if contrary views can be held on it without those views being contrary to reason" (Dearden, 1984, p. 86). Most people would agree that the science–religion issue is controversial both according to this definition and to the commonsense understanding of the term "controversial," though we can note that Dearden's definition is really a definition of "uncertainty." Controversy surely requires that some significance be attached to the uncertainty. For example, there is uncertainty about many matters in geography (the precise heights of mountains, the sources of rivers, etc.) but little or no controversy attaches to many (though not all) such uncertainties.

For someone to say that an issue is controversial is not to mean that they themselves necessarily consider it controversial—they may hold that the matter is completely uncontroversial—but that a range of views can be held on it by people without those views being unreasonable. So, for example, the teaching of sex education, whether we should have capital punishment and the significance of global climate change are all controversial whereas the value of π, the outcome of the Second World War, and whether slavery is desirable are not.

Furthermore, an issue may be controversial at one time but not at another—slavery was certainly controversial in the middle of the nineteenth century even if it is not so now (despite the fact that it still occurs in parts of the world), and I suspect that the significance of global climate change will become less controversial as the years go by—and it is easier for us today than it was in Dearden's time to see that a matter may be controversial for some people but not others. After all, the science-religion issue is not controversial, merely exasperating, for Richard Dawkins. A more general point is that controversy is often neither entirely absent nor completely present; there are degrees of controversy.

Importantly, a matter can be controversial without it being scientifically controversial. There are very few professional scientists for whom the young Earth hypothesis (i.e. that the Earth is some 10,000 or fewer years old—at any rate, orders of magnitude younger than the widely accepted figure among professional scientists of around 4,600 million years) is scientifically controversial: it is simply judged to be wrong. In the same way, there are very few professional scientists for whom the theory of evolution (in the sense that all life today is presumed to derive from extremely simple ancestors and ultimately from inorganic materials) is a controversial part of science. However, this last example illustrates the fact that a scientific theory may not be controversial even though aspects of it are. For example, precisely how life evolved from inorganic materials is still very unclear, as Leslie Jones points out in Chapter 12, and there is at least some degree of controversy about such matters as the relative importance attached to natural selection versus other agents of evolutionary change, how long it typically takes for a new species to arise and many, many details of the history of life—how one group of organisms evolved into another being chief among these, though we are learning more (e.g. about the origins of whales, bats and the earliest land animals) all the time.

Would one want explicitly to teach about creationism in science lessons? Both the knee-jerk and the considered reaction from most scientists and science educators has been "no." Here my interest is not in the legal situation in different countries—carefully discussed with reference to the USA by Randy Moore in Chapter 2—but in whether it would be desirable on educational grounds to teach about creationism in science lessons. Given the preceding paragraph, I would not want any such teaching, were it to occur, to give the impression that creationism and the theory of evolution are equally valid scientifically. They are not (and nor is it appropriate to insist on spending equal amounts of time on evolution and creationism in science lessons).

However, I do not belong to the camp that argues that creationism is necessarily nonscientific. For all that I have no doubt that the overwhelming majority of those who believe in creationism (and intelligent design theory) do so because of their religious beliefs, it is logically possible to hold that evolution (*sensu* major anatomical, physiological, genetic, and biochemical changes in organisms) has not happened. I can conceive of a world in which the earliest fossils indicate creatures as complicated as ourselves and in which the same geological strata show human and dinosaur footprints. Indeed, I very much favor students examining the evidence for evolution in a critical manner as advocated by Abdul Mabud in Chapter 7. I am more confident than Abdul Mabud that the evidence points towards the truth of the conventional scientific account of evolution, but he and I certainly agree on the need for students to exercise their critical faculties by examining the evidence.

Furthermore, I am not convinced that something being "nonscientific" is sufficient to disqualify it from being considered in a science lesson. An understanding of (nonscientific) context often helps in learning the content of science, partly by motivating the learner, partly by indicating why the content is meaningful and partly by rendering the content more intelligible. Indeed, the distinction between context and content can be more blurred in the biological and earth sciences than in physics and chemistry. When a student learns about the cellular mechanisms and genetic principles that underlie cystic fibrosis, is cystic fibrosis at least in part the content or merely the context for membrane transport and inheritance? As far as evolution is concerned, stories about, for example, the voyage of *The Beagle* or the domestication of pigeons can simply be motivating for students or can also help clarify the thinking that Darwin went through, thus enabling students themselves better to understand the science.

An additional point is that much of the debate about the teaching of evolution is still situated, albeit implicitly, in a view of education in which the teacher has complete control over that which enters each student's mind about the issue in question. Such a view has been outmoded for years. If one turns from considering that which is taught to that which is learned, it immediately becomes clear that students have always (and nowadays possibly more than ever) learned outside of formal lessons and from people (including family) and other sources in addition to their prescribed textbooks.

This leads onto the more general point about how one should teach about controversial issues in science. The short answer is that one should do so in a way that respects evidence and valid reasoning, that helps students to understand which aspects of an issue are controversial and which are not (e.g. with reference to global warming it is not controversial that atmospheric carbon dioxide levels are rising, there is a small, though decreasing, degree of controversy attached to the role of humans in this, and there is very considerable controversy as to the impact of this rise).

As to the precise techniques one can employ when teaching a controversial issue, there are many (e.g. Wellington, 1986). I have tried, in addition to other approaches (e.g. advocacy and affirmative neutrality), the approach of procedural neutrality (in which the teacher acts as a facilitator but does not try to steer the debate in any particular direction). Graduate science students doing a one-year course to become specialist science teachers were presented with a wide range of materials detailing how Darwin's *On the Origin of Species By Means of Natural Selection* was received after its publication on November 24, 1859. The students then divided themselves into four groups and desktop published the front page of a newspaper of that time reacting to Darwin's book. One group produced *The Times* that, though it discussed the book, gave pride of place to Garibaldi's campaign in Italy. Another group produced *Nature* which, though generally positive about the book, stressed the controversy caused in the scientific establishment. A third group produced *The Church Times* that gave pride of place to Lord Wilberforce's scientific assessment of Darwin's theory. Finally, the fourth group formed a feminist workers' cooperative and produced a tabloid called *The Splurge*. Their lead story was headed "All Men are Apes/ It's Official!!!"

Teaching about Matters of Personal Significance

Even if one agrees that there are aspects of the science-religion relationship that can or should validly be taught in science lessons, this is not to minimize the fact that teaching about this relationship is, for many, a matter of considerable personal significance. Indeed, the contributions in this book by Wolff-Michael Roth (Chapter 8) and David Jackson (Chapter 11) indicate some of the ways in which this is the case, whether for students or for staff.

Teaching about a matter of personal significance is related to, but not identical to, teaching about a controversial issue. To teach about matters of personal significance can demand much of us as teachers. It exposes aspects of ourselves to our students in a way which many teachers will find threatening or invasive, though some may find exciting. There are parallels to teaching about ourselves as sexual beings when teaching sex education—a dangerous state of affairs and one that I caution new entrants to the teaching profession to avoid, at the least for their own sakes.

Equally, teaching about matters of personal significance can make significant demands on students. We need to find ways that are respectful of students that neither threaten their beliefs nor molly coddle them as if getting them to think about their beliefs was necessarily to attack them. In England and Wales, a course for 16–19 year olds that I direct, titled Salters-Nuffield Advanced Biology, has been running in a pilot version since September 2002 and nationally since September 2005. One of the learning objectives in the topic on evolution and ecology is "Appreciate why, for cultural reasons, the theory of evolution has been so controversial for some people" (Edexcel, 2005, p. 31). The reason for including this learning objective is because the team devising the course felt it worthwhile for 16–19 year olds to understand something about this issue. Of course, the aim is neither to persuade students to embrace certain religious beliefs nor to cause them to abandon them: it is to help them learn and to clarify their thinking. The material in the student textbook that relates to this learning objective is quoted in Box 13.1 (below).

> **Why Is the Theory of Evolution Controversial?**
>
> Darwin had considered training to be a clergyman and knew full well that his ideas about evolution meant that he seemed to be challenging the Bible. The first two chapters of Genesis record how all of the Earth's organisms were created by God about 6,000 to 10,000 years ago, in much the form that they exist today.
>
> It is hard for many people to appreciate just how controversial Darwin's ideas were at the time. To this day there are many people in the world who cannot reconcile their religious beliefs and the theory of evolution. Few Muslims accept it and many Christians either reject or aren't comfortable with it either. In the UK around 10% of people are **creationists**. Creationists accept the literal teaching of the Bible, Qur'an or other scriptures and reject the theory of evolution. Creationists believe instead that God specially created Adam and Eve out of the dust of the earth so that humans are quite distinct from all other species. In the US around 40% of people are creationists. Indeed, worldwide there is no doubt that the idea that all organisms are descended from a common ancestor that lived some three thousand million years ago—which is what evolutionary biologists believe—is a minority position.
>
> The great majority of biologists, geologists and other scientists do accept a modernized version of evolution called **neo-Darwinism**. Neo-Darwinism combines Darwin's and Wallace's theory of natural selection with what we now know about inheritance. Many scientists who accept the theory of evolution in this modernized version also have firm religious beliefs. Such scientists have various ways of reconciling their religious beliefs with the theory of evolution. For example, they may believe that God created the original conditions that enabled the universe to come into existence, and then allowed evolution to take its course. In this understanding, God gives the whole of creation, including us, a certain freedom. Life is not predetermined but open ended.

Box 13.1 Extract from the Salters-Nuffield Advanced Level course (Hall *et al.*, 2006, p. 51)

Almost everyone reading Box 13.1 can probably find something in it that they don't like. However, our purpose in writing it was to avoid shying away from the science-religion issue and to attempt to present material that would not be considered unfair or disrespectful by readers, whatever their own positions on the issue. There is an optional activity that teachers have available to help them when teaching this part of the course. This activity is a role-play for students, and the student sheet that accompanies it is presented in Box 13.2.

Activity 5.16 The Controversial Nature of the Theory of Evolution

This activity gets you to engage in a discussion to help you understand why, for cultural reasons, the theory of evolution has been so controversial for some people. In groups you and your peers will engage in a modest role play in which you defend a range of positions with respect to the evolution–creationism debate. Each of you will have one of the four roles outlined below.

This may be an issue about which you have thought carefully or not at all. The aim of the activity is not necessarily to get you to change your own views but to appreciate how different people can hold different views on this issue.

Atheist with a strong belief in evolutionary biology

You have spent a reasonable amount of time thinking about religion and are certain that there is absolutely no truth in any religious claims. You are happy to describe yourself, if asked, as an atheist. You don't try and force your views on others but intensely dislike and consider wholly unjustified any attempt by someone with a religious faith to convert others to their point of view.

You think the world would be a better place without religions. What you particularly dislike is the way in which religious people seem certain that their view is right. You are extremely dubious about the extent to which anyone with a religious faith can be a decent scientist.

One of the exciting things about evolutionary biology, you feel, is the way in which it can explain, through natural selection, why people have a religious faith. You haven't had time to read the latest books on the subject but know that the authors argue that religions evolved to make people more likely to be altruistic and get along with their neighbours. This would have had survival value.

Agnostic

You haven't really given much thought as to whether there is a God or not and feel you have an open mind on the question. You rather wish more people were prepared to have an open mind about such matters as it seems to you that half the world's problems are caused by people being so certain about things of which they know rather little.

> ***Religious belief and belief in evolutionary biology***
> You have a strong religious faith and are pretty sure that the conventional evolutionary biology account is correct. It rather frustrates you that most people seem unable to read the scriptures in any way other than the more literal fashion. Whether or not miracles occurred in the past they don't seem to nowadays much, if at all, and, anyway, the important thing about the scriptures is not the miracles they describe but the way they point to the nature of God and to what it is to lead a flourishing life.
>
> You feel strongly that the combination of a strong religious faith and a belief in evolutionary biology is the best way to understand human nature. Evolutionary biology is concerned with factual questions of how life (including humans) evolved. Religion is concerned with ultimate questions of value and being.
>
> ***Creation scientist***
> You have a strong religious faith and accept the scientific method. However, you don't think that the scientific evidence points to an old Earth; quite the opposite. You feel that the scientific account that makes the most sense is that life arose on Earth round about 10,000 years ago and soon after was almost totally destroyed by a giant flood. What science cannot do is to say why these events occurred: this is provided by scripture.
>
> You think that it is a really good idea for advanced level biology students to examine critically the evidence for evolution and feel confident that if they are given the time to do this, many of them will realise that evolution is confined to natural selection within species or possibly leading to the evolution of closely related species over a period of a few thousand years.

Box 13.2 Extract from the Salters-Nuffield Advanced Level course (www.snabonline.com/ accessed 16 April 2006; password protected).

Part of the hope would be, as it often is with role plays, that undertaking this activity, and the subsequent classroom discussion, would help students to better understand one another and that this understanding might help them to respect views other than their own.

Conclusion

The strongest argument as to why science teachers should deal with the relationship between science and religion when teaching about origins, whether in biology, earth sciences, or astronomy, is that it is good science teaching to do so. As Lee Meadows points out in Chapter 10, it is unlikely that teaching in this area can help students for whom there is a conflict between science and their religious beliefs to resolve the conflict, but good teaching can help them to manage the conflict—and to learn more science. We can help students to find their science lessons interesting and intellectually challenging without their being threatening. Effective teaching in this area can not only help students learn about the theory of evolution but also better appreciate the way science is done, the procedures by which scientific knowledge accumulates, the limitations of science and the ways in which scientific knowledge differs from other forms of knowledge.

References

Barbour, I. G. (1990). *Religion in an age of science: The Gifford lectures 1989-1991, Volume 1.* London: SCM.
Brooke, J. H. (1991). *Science and religion: Some historical perspectives.* Cambridge: Cambridge University Press.
Chalmers, A. F. (1999). *What is this thing called science? (3rd ed.)* Buckingham, England: Open University Press.
Dearden, R. F. (1984). *Theory and practice in education*, London: Routledge & Kegan Paul.
Edexcel (2005). *AS/A GCE in Biology (Salters-Nuffield): Specification.* Available at http://www.edexcel.org.uk/VirtualContent/72372/Biology__Salters_Nuffield__8048_9048_spec.pdf (accessed 16 April 2006).
Gould, S. J. (1999). *Rocks of ages: Science and religion in the fullness of life.* New York: Ballantine.
Hall, A., Reiss, M., Rowell, C. & Scott, A. (Eds.) (2006). *Salters-Nuffield Advanced Biology A2 Student Book.* Oxford: Heinemann.
Kuhn, T. S. (1970). *The Structure of scientific revolutions (2nd ed.).* University of Chicago: Chicago Press.
Marples, R. (Ed.) (1999). *The aims of education.* London: Routledge.
Merton, R. K. (1973). *The sociology of science: Theoretical and empirical investigations.* University of Chicago Press, Chicago.
Popper, K. R. (1934/1972). *The Logic of Scientific Discovery.* London: Hutchinson.
Reiss, M. J. (1992). How should science teachers teach the relationship between science and religion? *School Science Review, 74*(267), 126-130.
Reiss, M. (2005). The nature of science. In J. Frost & T. Turner (Eds), *Learning to teach science in the secondary school: A companion to school experience, 2nd ed.* (pp. 44-53). London: Routledge Falmer.
Reiss, M. J. (2007). What should be the aim(s) of school science education? In D. Corrigan, J. Dillon & R. Gunstone (Eds.), (pp. 13-28). *The re-emergence of values in the science curriculum*, Rotterdam: Sense.
Wellington, J. J. (Ed.) (1986). *Controversial issues in the curriculum.* Oxford: Basil Blackwell.

Notes on Contributors

Marilyn Bechler is an Art Educator whose own work employs a variety of media including: ink and pencil drawings, photography, printmaking, jewelry and silversmithing, and art glass. She has a BA in Fine Arts from Indiana University and an MAE from Valdosta State University. She is currently the art teacher at Pine Grove Elementary School in Valdosta, Georgia. She regularly uses aspects of science as the focus of art lessons because she sees the natural world as a work of art that lends itself as an inspiring subject to study. She has shared her expertise with the Biology Department at Valdosta State University by co-teaching Gyotaku (Japanese Fish Printing) in Ichthyology classes and developing an ecology drawing activity, "Biome Critter Art," that is part of designated science content courses for education majors. After residing in Georgia for the last twelve years, she is well aware of the challenges of teaching evolution to creationist students and has supported the recent inclusion of evolution in the new state curriculum, the Georgia Performance Standards.

David L. Haury is an Associate Professor of science education at The Ohio State University in the College of Education and Human Ecology. He teaches doctoral seminars focusing on research in science, mathematics, and technology education, and he offers courses on instructional methods in science for prospective and practicing teachers. Having an interest in evolution since childhood, David has been an active advocate of evolution education in schools, churches, and teacher preparation programs since the 1970's. He has assisted Ohio's State Board of Education and the Ohio Department of Education in responding to creationist initiatives and supporting the teaching of evolution. At his university, he participated with colleagues in the sciences to develop an Interdisciplinary Evolutionary Studies program for undergraduate students. David is the former Director of the ERIC Clearinghouse for Science, Mathematics, and Environmental Education, and the founding editor of the Journal of Science Teacher Education.
http://ehe.osu.edu/edtl/faculty/HauryDavid.htm
http://www.lifeevolving.org/

David F. Jackson is an Associate Professor of Science Education and Graduate Coordinator in the Mathematics and Science Education Department at the University of Georgia, USA. His primary teaching focus is on the education, with regard to both science and pedagogy, of middle school science teachers, with a special emphasis on the history of the Earth and of life. He formerly taught general, Earth, and physical science in grades 5, 7, and 9 in independent schools in Minnesota and Pennsylvania and Computer Applications in the Detroit Public Schools. Before teaching he gained research experience in

invertebrate paleontology as an assistant to Stephen Jay Gould at the Museum of Comparative Zoology at Harvard and as a summer fellow under Niles Eldredge at the American Museum of Natural History. The interaction of science and religion was a strong personal interest from a very early age, but was brought to the forefront of his professional concerns by the variety of creationist views held by many of his science education students at Georgia. He holds an A.B. in Geological Sciences, *cum laude* in general studies, from Harvard University, and an Ed.D. in Science Education and Instructional Technology from the University of Michigan.
http://www.coe.uga.edu/mse/faculty/jackson/index.html

Leslie S. Jones is an Assistant Professor in the Department of Biology at Valdosta State University. She teaches standards-based, inquiry-oriented science content courses for prospective and practicing teachers, as well as several courses based on evolution to upper level biology majors and non-majors. Since Valdosta is in rural South Georgia, most of her current students have strong creationist beliefs and know very little about biological evolution. These classes serve as sites for practitioner/action research initiatives concentrated on improving postsecondary science pedagogy and addressing educational equity issues. Education is the second focus of her scientific career and follows years of active research in equine reproductive physiology and cell biology. She has published basic science research in journals such as *Molecular and Cellular Endocrinology, Biology of Reproduction, The Journal of Reproduction and Fertility, and The Journal of Animal Science.*
http://www.valdosta.edu/biology/jones.htm.

Shaikh Abdul Mabud is Director General of the Islamic Academy in Cambridge where he has been based since 1983. He is also Professor of Islamic Philosophy at the Islamic College for Advanced Studies, which is affiliated with Middlesex University, London. His work at the Academy includes curriculum designing, teacher training, and research in moral and spiritual education, and the philosophy of science education. He is Editor of Cambridge-based *Muslim Education Quarterly*, and Corresponding Editor of *PANORAMA: International Journal of Comparative Religious Education and Values* published from Germany. Alongside various other memberships, he is also a Fellow of The Royal Society for the encouragement of Arts, Manufactures and Commerce, UK (FRSA).He has published numerous articles in both scientific and educational journals. Books of his include *Sex Education and Religion* (co-editor Michael Reiss), 1998, The Islamic Academy, and *Theory of Evolution: An Assessment from the Islamic Point of View*, 1994, The Islamic Academy. Prior to his joining the Islamic Academy, he taught physics at the University of Rajshahi and the University of Utah. He holds a PhD in Physics and a PGCE in science from the University of Cambridge.

NOTES ON CONTRIBUTORS

Lee Meadows is an Associate Professor in the Department of Curriculum and Instruction at the University of Alabama at Birmingham, USA. His work as a methods professor and school reformer focuses on the implementation of inquiry-based teaching in secondary science classrooms. He has published previously on the teaching of evolution to religious fundamentalists, including articles in *The Journal of Research in Science Teaching* and *The American Biology Teacher*. He grew up in the fundamentalist faith of which he writes. Those experiences have given him a goal of helping to delineate a pathway by which religious students can find success in the science classroom without feeling that they must abandon their faith. He serves as an elder in a Presbyterian church.

David Mercer is an Associate Professor in Science and Technology Studies at the University of Wollongong, Australia. He has published on law and science, scientific controversy, public understanding of science and the history of communication technology. Between 2001-2004 he served as a member of the National Committee for History and Philosophy of Science of the Australian Academy of Science and is a member of the editorial board of Metascience. Publications relevant to creation science debates include: Gary Edmond & David Mercer, 'Anti-social Epistemologies', *Social Studies of Science* December (2006) vol. 36, no. 6, pp. 843-853 and Gary Edmond & David Mercer, 'Creating Science: Science, Law and Religion in the Australian Noah's Ark Case' (1999) 8 *Public Understanding of Science*, 317-343.

Randy Moore is the H.T. Morse-Alumni Distinguished Teaching Professor of Biology at the University of Minnesota. Randy has written several biology textbooks and laboratory manuals, and edited *The American Biology Teacher* for 20 years. In addition to teaching introductory biology and courses about the evolution-creationism controversy, he has written a variety of books about evolution, including In the Light of Evolution: Science Education on Trial, Evolution 101, and the upcoming With Flying Banners and Beating Drums: The People and Places of the Evolution-Creationism Controversy. Randy speaks about the evolution-creationism controversy throughout the United States.

Robert T. Pennock is Professor at Michigan State University, where he is on the faculty of the Lyman Briggs School of Science, the Philosophy Department, the Department of Computer Science, and the Ecology, Evolutionary Biology and Behavior graduate program. His research interests are in philosophy of biology, evolutionary computation, and in the relationship of epistemic and ethical values in science. Some of his scientific research on experimental evolution and evolutionary computation was featured in a cover story in *Discover* magazine. He speaks regularly around the country on issues of science and values, and was named a national Distinguished Lecturer by Sigma Xi, The

Scientific Research Society. His book *Tower of Babel: The Evidence against the New Creationism* has been reviewed in over fifty publications; the New York Review of Books called it "the best book on creationism in all its guises." He was an expert witness in the historic *Kitzmiller v. Dover Area School Board* case that ruled that Intelligent Design is not science and that teaching it in the public schools is unconstitutional. A Fellow of the American Association for the Advancement of Science, he serves on the AAAS Committee on the Public Understanding of Science and Technology.

Michael Poole is Visiting Research Fellow in Science and Religion in the Department of Education and Professional Studies at King's College, London. He taught Physics and some Religious Education at a boys comprehensive school in London, where he was Head of the Physics Department. This was followed by three years preparing and broadcasting radio programs on science-and-religion for overseas audiences, together with part-time lecturing in science education at King's College. He joined the full-time staff in 1973 as Lecturer in Science Education (Physics). His research interests are in the interplay between science and religion, with special reference to its educational context. He is the author of several books, including: *A Guide to Science and Belief*, (Oxford: Lion, 1990, 1994 & 1997) [8 languages], *Beliefs and Values in Science Education*, (Buckingham: Open University Press, 1995) [2 languages], joint editor and contributor to *God, Humanity and the Cosmos: a Textbook in Science and Religion*, Edinburgh: T & T Clark, 1999, 2003 & 2005 and *User's Guide to Science and Belief*, Oxford: Lion Hudson, 2007 and some seventy papers and articles on science and religion for science teachers, RE specialists and a general readership. http://www.kcl.ac.uk/ schools/sspp/education/staff/mpoole.html

Michael J. Reiss is Professor of Science Education at the Institute of Education, University of London and Director of Education at the Royal Society. After a Ph.D. and post-doctoral research in animal behavior and population genetics, he taught in schools before returning to the university sector. He is a Priest in the Church of England, Chief Executive of the Science Learning Centre, London, Honorary Visiting Professor at the University of York, Docent at the University of Helsinki, Director of the Salters-Nuffield Advanced Biology Project, a member of the Farm Animal Welfare Council and editor of the journal *Sex Education*. Books of his include Braund, M. & Reiss, M. J. (Eds) (2004) *Learning Science Outside the Classroom*, RoutledgeFalmer, Levinson, R. & Reiss, M. J. (Eds) (2003) *Key Issues in Bioethics: A Guide for Teachers*, RoutledgeFalmer and Reiss, M. J. (2000) *Understanding Science Lessons: Five Years of Science Teaching*, Open University Press.
www.reiss.tc

NOTES ON CONTRIBUTORS

Wolff-Michael Roth is Lansdowne Professor of Applied Cognitive Science at the University of Victoria, British Columbia, Canada. From 1992 on, already working at the university, he taught science in British Columbia elementary schools at the fourth-seventh grade levels, always associated with research on knowing and learning. More recently, he has conducted several ethnographic studies of scientific research, a variety of workplaces, and environmental activist movements. Wolff-Michael Roth publishes widely and in different disciplines, including linguistics, social studies of science, and different subfields in education (curriculum, mathematics education, science education). His recent books with SensePublishers include *Doing Qualitative Research: Praxis of Method* (2005), *Learning Science: A Singular Plural Perspective* (2006), and, with Ken Tobin, *Teaching to Learn: A View from the Classroom* (2006). He also edited *Auto/Biography and Auto/Ethnography: Praxis of Research Method* (2005) and, co-edited, with Ken Tobin, *The Culture of Science Education: A History in Person* (2007) and *Science, Learning, Identity: Sociocultural and Cultural-Historical Perspectives* (2007).

Michael Ruse is a professor of philosophy at Florida State University. He is the author of many books on Darwinism and the Evolution-Creation controversy, including most recently *The Evolution-Creation Struggle* published by Harvard University Press. In 1981, he was a witness for the ACLU in a successful court case in Arkansas against the teaching of Creation Science in publicly supported schools, a battle he documents in the edited volume *But is it Science?* published by Prometheus Press.

Index

Abrahamic religions, 77
American Civil Liberties Union, 14
Anthropic Cosmological Principle, 78, 82
Anglicanism, 2, 34–35, 39–40
Arjuna, 3
Arkansas Baptist Convention, 13
Astronomy, 10, 163, 207
Atheism, 3–4, 12, 37–41, 83, 112, 161, 187
Augustine, 79
Australian Science Education Project, 138
Ayres Rock, (See Uluru)

Barbour, Ian, 5–6, 168, 201
Behe, Michael, 26, 62, 72, 101
Benchmarks for Science Literacy, 126, 165, 171
Bible Belt, 182
Big Bang, 59, 75, 78, 113, 116, 159, 161, 167, 169, 172
Biological Sciences Curriculum Study, 16, 19
Bishop of Oxford, 32, 35
Bryan, William Jennings, 12–13, 51
Buddhism, 4
Burning Bush, 3

Carter, Jimmy, 162
Caucasoid, 128
Christianity, 4, 8, 11, 15, 28, 31–41, 75–76, 85, 169, 177, 193
Church of England, 8, 31–35, 212
Communism, 86
Comte, Auguste, 45

Darrow, Clarence, 15, 51
Darwin, Charles, 7, 11, 31–33, 51, 109
Darwin, Erasmus, 32
Dawkins, Richard, 5, 8, 38–40, 46–47, 202
Deep time, 159, 161, 165, 169
Dembski, William, 27, 62
Descartes, Rene, 7

DeWolf, David, 59, 61
Dharma, 4
Discovery Institute, 26, 59–64, 66–68, 71–72
DNA, ix, 87, 98, 126, 128, 130, 132, 134, 190–191
DNA fingerprinting, 132
Dobzhansky, Theodosius, 36

Earth Sciences, 9–10, 78, 166–167, 172, 203, 207
Eastern Orthodox, 4
Edwards v. Aguillard, 20, 22—24, 63, 68
Eohippus, 97
Episcopal, 2, 168, 183
Epistemology, 8, 43, 48, 51, 53, 55, 105–106, 110, 120, 194
Eugenics, 56
Extinction, 27, 135–136
Euthanasia, 5
Exodus, 3

Federal Trade Practices Act, 56
Flood geology, 16, 75–77
Fossils, ix, 39, 64, 90–92, 126, 129, 131, 160, 166–168, 174, 191, 203
Fossil record, 16, 91–92, 95–97, 126–127, 165, 173, 191

Galapagos Islands, 33, 91, 190
Galileo, 7, 31, 50, 79, 87, 125, 194
Genesis, ix, 2, 16, 20–21, 26, 75, 77–79, 98, 147, 169, 183, 188, 204
Genetic modification, 45
Geologic timescale, 59
Global warming, 27, 127, 203
Golden Age of Islam, 83
Gould, John, 190
Gould, Stephen Jay, 8, 38, 46–47, 68, 145, 210
Gray, Asa, 11, 35, 80

INDEX

H.M.S. Beagle, 33, 190, 203
Haeckel, Ernst, 64
Hajj, 4
Ham, Ken, 28
Hebrew scriptures, 1, 77, 183
Holy Spirit, 4
Hominids, 129, 131, 189
Homo erectus, 132
Homo floresiensis, 133
Homo habilis, 132
Homo neanderthalensis, 133
Homo sapiens, 137
Huxley, Julian, 86
Huxley, Thomas Henry, 11, 26, 31, 35, 40, 81, 86, 87

Inherit the Wind, 17, 51
Institute for Creation Research, 67
Intelligent Design, 2–3, 8, 25–26, 37, 48–49, 55, 59–74, 82, 129–130, 138, 164, 184, 193, 203, 211
Irreducible complexity, 82
Islam, 2, 4, 9, 83, 85, 122, 147, 169, 177, 201, 205, 210

Johnson, Phillip, 26–27, 60
Judaism, 85
Judeo-Christian, 2–3, 6, 100, 194

Kingsley, Charles, 81–82
Kitzmiller v. Dover, 60, 72
Kuhn, Thomas, 199-200

Lamarck, Jean Baptiste de, 32, 37
Laudan, Larry, 48, 53
Leibniz, 7
Lewis, C.S., 169
Lewontin, Richard, 148
Linnaeus, 191
Logical positivism, 84
Lutheran, 2, 183
Lyell, Charles, 33, 77

Macroevolution, 65, 100–101, 103, 127, 165–166
Malthus, Thomas Robert, 34
Marxism, 49
McCarthy Era, 17
Meditation, 3, 5
Merton, Robert, 198–199
Metaphysical, 6, 45, 47, 84–86, 125, 129
Microevolution, 65, 89, 99–101, 165
Miller-Urey, 192
Minnich, Scott, 61, 72
Modernism, 4, 11
Monkey Trial, 14, 51
Morris, Henry, 16, 21, 77
Moses, 3
Mount Fuji, 4
Muslim (See Islam)
Myer, Stephen, 61–62

National Center for Science Education, 46, 66–67
National Science Education Standards, 126
Natural selection 11, 26, 31–35, 38, 44, 66–67, 80–81, 85, 87, 95–96, 99–100, 102, 133, 139, 167, 190, 193, 202, 204–207
Naturalism, 84
Nazism, 86
Neanderthals, 127
No Child Left Behind Act, 63, 69
Non-overlapping magisterial, 46
Northern Ireland, 40

Oklahoma Baptist Convention, 13
Old Testament, (See Hebrew Scriptures)
On the Origin of Species, 11, 18, 31–32, 34, 99, 204
Origin of life, x, 1–2, 45, 67, 163, 168, 188, 191–192

Pagan, 15, 41
Paleoanthropology, 97, 127–128, 133
Paley, William, 33, 86
Pedagogy, 9, 71, 133, 209–210
Peppered moth, 99–100, 165
Philosophie Zoologique, 32

Piltdown Man, 64
Pope John Paul II, 2, 168, 184
Pope Pius XII, 184
Popper, Karl, 48, 199
Population genetics, 8, 212
Postmodernism, 52, 173
Prayer, 3-4, 17, 21–22, 115, 146–147
Presbyterian, 2, 62, 168, 183, 211
Price, George McCready, 77
Principles of Geology, 33, 77

Quaker, 39–40
Quran, 2, 37, 184

Reification, 85
Roman Catholic Church, 2, 37, 168, 183-185
Russell, Bertrand, 84

Salat, 4
Sangha, 4
Scopes Monkey Trial, 14, 17
Scopes, John, 14, 18, 51
Scott, Eugenie, 3, 46, 67
Sectarianism, 44
Seventh Day Adventists, 77
Shahada, 4
Spencer, Herbert, 44
Sputnik I, 16

Teach the controversy, 59–62, 65, 70, 72–73
Ten Commandments, 4, 21, 68
Theologians, 4, 7, 12, 34, 77
Theology, 6, 38, 46, 76, 172, 181, 187
Torah, 4

Uluru, 4
Umma, 4

Wedge strategy, 61
Wells, Jonathan, 26, 60, 62, 64
Whitcomb, John, Jr., 16, 77
White, Ellen, 77
Wilberforce, Samuel, 31

Worldview, 2–6, 10, 129, 137–138, 147–151, 155–156, 159, 169, 172, 193, 201

Zakat, 4
Zygon, 5

Studies in the Postmodern Theory of Education

General Editors
Joe L. Kincheloe & Shirley R. Steinberg

Counterpoints publishes the most compelling and imaginative books being written in education today. Grounded on the theoretical advances in criticism, feminism, and postmodernism in the last two decades of the twentieth century, Counterpoints engages the meaning of these innovations in various forms of educational expression. Committed to the proposition that theoretical literature should be accessible to a variety of audiences, the series insists that its authors avoid esoteric and jargonistic languages that transform educational scholarship into an elite discourse for the initiated. Scholarly work matters only to the degree it affects consciousness and practice at multiple sites. Counterpoints' editorial policy is based on these principles and the ability of scholars to break new ground, to open new conversations, to go where educators have never gone before.

For additional information about this series or for the submission of manuscripts, please contact:

>Joe L. Kincheloe & Shirley R. Steinberg
>c/o Peter Lang Publishing, Inc.
>29 Broadway, 18th floor
>New York, New York 10006

To order other books in this series, please contact our Customer Service Department:

>(800) 770-LANG (within the U.S.)
>(212) 647-7706 (outside the U.S.)
>(212) 647-7707 FAX

Or browse online by series:
>www.peterlang.com